Forests, Peasants, and Revolutionaries:

Forest Conservation and Organization in Soviet Russia, 1917-1929

Brian Bonhomme

East European Monographs, Boulder
Distributed by Columbia University Press, New York
2005

EAST EUROPEAN MONOGRAPHS, NO. DCLIV

ISBN: 0-88033-553-X

Library of Congress Control Number: 2005921127

Printed in the United States of America

Table of Contents

I
Introduction

The emergence of forest conservation as a significant public concern in Russia is best dated from the last quarter of the nineteenth century, about three decades prior to the early Soviet period on which this book is primarily focused. The concentration of much of the Russian Empire's population in European Russia, rather than in the far north or across the huge expanses of Siberia and the Far East, had for centuries mitigated pressure on the forests of those latter areas, and would do so well into the twentieth century. In European Russian, on the other hand, things were different. The clearing of land for agriculture, as well as growing industrial use, took an increasing toll during the decades immediately following the serf emancipation of 1861. Domestic use by the peasant population (for whom wood was the main source of fuel and building materials) added mightily to the pressure. Drawing for their inspiration on American, French, German, and other foreign examples, as well as significantly on their own traditions, Russian foresters and the Imperial government came together during these years to lay the foundations of a conservation regime whose main achievement was a fairly comprehensive—though not hugely effective—forest protection law passed in 1888 under the title "Statute on the Safeguarding of Forests" (*Polozhenie o sberezhenii lesov*; see chapter 2). Anxiety for the fate of Russia's forests, however, only increased thereafter. Calls for further and more drastic conservation measures became ever clearer and louder during the years leading up to the First World War. Though the war itself put forestry reform on hold, the revolutions of February and October 1917 (Old Style) opened up radical new possibilities. The Provisional Government, though not destined to enjoy power for long, sought to reform forest use and improve protection partly by granting increased controls over forest resources to local organs. The Soviet regime that seized power in October—and held it thereafter—took even greater interest in the forests, including in conservation, which it claimed to understand—and set out to establish—as an integral part of the new socialist order.

The raison d'être of Soviet conservation was from the start twofold. Economic considerations were primary. Forest use was to be planned and managed along the most scientifically—and technically—advanced lines so as to achieve maximal yields in efficient and sustainable manner. But at the same time, a social goal was implicit: to enroll the peasant masses—those who lived and worked in, around, and off the forests—in the new socialist order. As in other areas of life, so in forest matters, this involved inculcating collectivist-socialist rather than individualist or village-based attitudes and loyalties in large part through the establishment and maintenance of centrally-planned systems for resource management and use. Accordingly, major forest codes—the "Basic Law on Forests" (*Osnovnoi zakon o lesakh*) and the "Forest Code" (*Lesnoi kodeks*)—were passed in 1918 and 1923, respectively. These are treated in detail in subsequent chapters.

The first decade of Soviet forest conservation has been little studied, especially by English-language scholars. Brenton Barr and Kathleen Braden's *The Disappearing Russian Forest* surveys a broad period and consequently moves rather quickly and lightly over these early years.[1] Other, more general works on Russian and Soviet resource conservation also tend to give relatively short shrift to the revolutionary years, preferring to emphasize developments from the 1960s forward. The same is largely true also of Post-Soviet Russian-language studies. A recent Russian work, translated as *A History of Russian Forestry and Its Leaders*, also paints a broader picture.[2] Soviet-era writings, on the other hand, often do place greater emphasis on the revolutionary period, and in particular on the figure of Lenin. However, in most cases they are heavily ideological and of limited scholarly use (see chapter 8, in particular).

Particularly relevant to the current study is Douglas R. Weiner's *Models of Nature: Ecology, Conservation and Cultural Revolution in Soviet Russia*. This treats early Soviet conservation generally and only occasionally offers direct discussion of forestry. Nonetheless, since its first publication in 1988 it has become the authoritative work on its subject and helped frame some of the main questions this study explores. In his introductory essay, for example, Weiner wrote: "through the early 1930s the Soviet Union was on the cutting edge of conservation theory and practice."[3] In fact, this claim—which has been seconded

---

[1] Brenton M. Barr and Kathleen Braden, *The Disappearing Russian Forest: A Dilemma in Soviet Resource Management* (London: Hutchinson; and Totowa, NJ: Rowman and Littlefield, 1988).

[2] V. K. Teplyakov, et al., *A History of Russian Forestry and its Leaders* (Pullman, WA.: Washington State University Press, 1998).

[3] Douglas R. Weiner, *Models of Nature: Ecology, Conservation, and Cultural Revolution in Soviet Russia* (Bloomington: Indiana University Press, 1988, x).

or cited by various other scholars[4]—is not based on a comprehensive study of early Soviet resource management policies, but instead derives mainly from an evaluation of a network of unique nature preserves (*zapovedniki*) established by the Soviet government during the periods of War Communism and New Economic Policy.[5] The preserves' uniqueness lay both in the unprecedented level of inviolability they envisioned—disallowing in theory all economic activity, including hunting, fishing, and the gathering of materials—and in the emphasis given to their role as "natural laboratories," areas in which proto-ecological scientific work would be conducted into the structure and dynamics of natural communities.[6]

The *zapovednik* project's historical importance has been well established and will not be challenged here. On the other hand, it should be immediately clear that nature preserves represent only one aspect of Soviet nature conservation overall and a tiny fraction of the total land area of the early Soviet state. How, one wonders, were natural resources being exploited or protected *outside* these reserves, across the vast majority of Russian lands? Moreover, if *zapovednik*-style conservation was possible or desirable in certain relatively remote, economically unimportant, or hard-to-exploit regions, what was the situation when conservation and exploitation imperatives more obviously clashed, as they must have done, say, regarding vital economic resources—coal, timber, furs, and so on—on a national scale? Put alternatively, how did conservation fit more generally into the broader structure of early Soviet socialist construction, especially given the Bolsheviks' overarching drive toward electrification and industrialization? It is with exactly these kinds of questions that this book is concerned.

---

[4] See, for example, Peter J. Bowler, *The Norton History of the Environmental Sciences* (New York and London: W. W. Norton, 1993), 513; Loren R. Graham, *Science in Russia and the Soviet Union: A Short History* (New York: Cambridge University Press, 1993, 243.

[5] The War Communist period lasted from 1918-1921. The term reflects the Bolsheviks' rapid implementation of communist economic relations, including grain requisitioning, in order to help fight the Civil War of the same years. The New Economic Policy of 1921-1928 marked a sharp shift in policy, allowing the limited re-emergence of small-scale private enterprise and the abolition of grain requisitioning in favor of a tax. The years 1921-1928 are usually considered a liberal era in comparison with what came before and after.

[6] In fact, violations of the "inviolability" of *zapovednik*—both by the general public and the state—seem to have been quite common. See chapters 5 and 9.

In search of answers, my original plan was to be fairly comprehensive, exploring treatment of wildlife, forests, fresh and saltwater resources, air quality, and so on. But the libraries and archives of Moscow and Saint Petersburg yielded so much material that this approach was quickly jettisoned in favor of a focus solely on forests. The forests, it turned out, made an excellent study. More than most natural resources, they covered wide expanses of Russian territory, thus providing exactly the desired contrast with the much smaller amount of land contained in the *zapovedniki*. At the same time, forests—unlike, say, wild birds (which were a central focus of conservation battles in Britain and elsewhere at a similar time)—constituted a clearly vital resource within the national economy to state and public alike. Thus they could not simply be locked up or their use regulated without much conflict.[7] A study of the forests, for this reason at least, must of necessity intersect constantly and deeply with a wide range of critical social and political issues central to early Soviet history overall. Rescued thus from the possibility of writing on an isolated and obscure topic, I simultaneously found myself dealing with well-worn historiographical categories—War Communism, the NEP, and so on—though bringing to these in the process, I hope—and especially to the latter—important new light.

The twin focus points of this book are two Soviet forest-law packages. The first, "The Basic Law on Forests" (*Osnovnoi zakon o lesakh*) was passed in May 1918. Situated politically and conceptually in the domain of War Communism, it represented a highly ambitious effort to inventory, allocate, manage, and reproduce forest resources in the most efficient, scientific, and organized manner possible. In the event, it fell hopelessly short, victim of a variety of circumstantial stumbling blocks and also of its own internal shortcomings. Chapters 4 through 7 tell its story. It was succeeded in 1923 by a new law, "The Forest Code" (see chapters 7, 9-10). This sought to apply lessons learned from the previous law, but also reflected the wider changes taking the Soviet state out of an era of revolution and civil war into the calmer times of the New Economic Policy, when heavy

[7] Exactly this type of absence of conflict is explored in Douglas Weiner's *A Little Corner of Freedom: Russian Nature Protection from Stalin to Gorbachev* (Berkeley, CA: University of California Press, 1999). Here, the argument is developed that through the Stalin years and beyond, conservationists were frequently viewed by Party and state as harmless eccentrics, individuals and groups whose plans and activities had tolerably little connection to vital economic, social, or political priorities, and who could thus be allowed to operate in relative freedom within an otherwise heavily repressive state. The individuals and groups involved in early Soviet forestry, it will be become apparent, enjoyed no such invisibility.

government intervention in the economy was mitigated by the emplacement of modest free-market reforms.

At the same time, however, considerable attention is devoted to a wider time period as well. The late nineteenth and pre-Revolutionary twentieth centuries in particular are assessed in some detail as indispensable background to the main events discussed, and in order that these might sit in an appropriate context. In this regard, an effort has been made to avoid (or correct) a tendency among historians (especially Soviet ones) to dismiss the tsarist period as of little interest in terms of conservation. The study ends in 1929, which is a useful stopping point for two reasons. First, it allows sufficient time to trace the results – and therefore usefully to assess—the 1923 law. Second, the beginning in earnest that year of the Stalin Revolution—especially the collectivization of agriculture—effectively took forest matters off in a new direction, bringing to an end a somewhat coherent "chapter" spanning the years 1917-1929.

If the details are at times complex, the basic conclusions are generally straightforward. First, it will be argued that early Soviet forest conservation, though it introduced some important innovations, was by no means an absolute new beginning. Rather, it drew heavily on pre-Revolutionary models, the value and achievement of which have hitherto been underrated or ignored. Second, in terms of actual results, early Soviet forest conservation appears to have fallen far short of the high goals it set itself and, in contrast to the *zapovednik* project, does not seem to warrant being judged "cutting edge" (unless, of course, that term be taken quite literally!). Third, and most important, Soviet efforts at forest protection during this period, it turns out, were significantly hampered—even undermined—not only by the inevitable temporary challenges of the Revolution and Civil War, but more fundamentally by the unfailing non-cooperation of the Russian peasantry, the nation's majority population, for whom nature conservation and forest planning held little appeal and were instead constant sources of irritation and resentment. This conflict, aptly characterized by a Bolshevik observer in 1924 as "the war between the population and the forest economy"[8] is the dominant theme of much of this study. In this context the transition from "On Forests" (1918) to the "Forest Code" (1923) is presented as a pointed reduction of original Bolshevik ambitions for a national and

[8] RSFSR. *Vserossiiskii tsentral'nyi ispolnitel'nyi komitet,* XI sozyv. Tret'ia sessiia. Zasedanie vos'moe (15 October, 1924), 368.

scientifically-planned forest economy in the face of indefatigable and widespread peasant resistance.

Other scholars and scholarly traditions are engaged in this study also, explicitly or otherwise. Prominent among these is Soviet scholarship. Here, there is relatively little diversity. In conservation as in all else, the Revolution was presented as a fundamental break with the past, replacing the tsars' few scattered and ineffective measures against "predatory" forest practices with a coherent and fairly effective system of planning, more efficient use, and better protection.[9] In this tradition the figure of Lenin looms especially large. To the great Bolshevik leader—especially in his capacity as signer[10] of "On Forests"—was assigned almost divine wisdom and foresight, tireless energy, unbending commitment, and marked success in the service of forest (and other) conservation. It is in dialogue with Soviet scholarship of this nature that chapters 1 and 2, focused on the pre-Bolshevik period, and chapter 8, treating Lenin, are partly conceived.

The ideas of V. E. Boreiko, a Ukrainian specialist who has published several books and articles touching on various aspects of Soviet conservation, have also been influential. Boreiko is for the most part strongly critical of Soviet approaches and achievements—on rare occasions to the point of engaging in sheer polemics. For example, in a long essay on forest abuses covering several centuries up to the present day, he dismisses "On Forests," the Bolsheviks' first major forest code, off-hand in a few lines as "not having operated, even for a day."[11] If this is a refreshing antidote to endless Soviet praises of the law (none of which offers any real evidence of its effectiveness), it is hardly more helpful.[12]

---

[9] One Soviet account (and hardly even among the most uncritically pro-Soviet ones) states, for example, that tsarist-era conservation "was realized only in the form of hunting laws and in special conservation [measures] for bison." Forest conservation, by this reckoning was non-existent. This is far from the truth. See G. A. Bogoslavskii, "Iz istorii okhrany prirody v pervye gody sovetskoi vlasti." In Stroitel'stvo Sovetskogo gosudarstva, *Sbornik statei k 70-letiiu doktora istoricheskykh nauk professora E. B. Genkinoi*. Izdatel'stvo "Nauka," Moscow, 1972, 259.

[10] Many Soviet sources imply in this connection that Lenin personally wrote or shaped "On Forests." In fact, it is unclear if he had any real input to it at all. (See chapter 8).

[11] V. E. Boreiko, "Plakala Sasha kak les vyrubili." In Boreiko, *Belye piatna istorii prirodookhrany SSSR, Rossia, Ukraina, tom vtoroi*. Seriia: Istoria okhrana prirody, vypusk 7 (Kiev: Kievskii ekologo-kul'turnyi tsentr, Tsentr okhrany dikoi prirody SoES, 1996), 29.

[12] Ibid.

Not only *did* the law operate (admittedly in very limited form), it provides a revealing window into early Soviet attitudes toward forest use, and was for several years a flashpoint of the forest tensions already noted between state and society. For these reasons at least, it occupies center stage here.

Mention could be made of various other writers, all of whom have in some way at least indirectly influenced this work. The studies of Joan Debardeleben and Marshall Goldman, respectively, provided early insights into the theory and practice of Soviet conservation, making grist for the mills of later writers including Weiner, Philip Pryde, John Massey Stewart, and many others.[13] Russian-speakers such as Mikhail Lemeshev, Ionas Bergkholtsas, and others have also made important contributions to debate on Soviet conservation.[14] Overall, however, the focus of almost all of these writers is the recent past. The early years, with which we are concerned here, remain a lightly-trodden field.

* * *

In researching and writing this book—and the doctoral dissertation from which it emerged—I have accumulated a considerable debt of gratitude which I can only begin to repay here. The History Departments of the City College and of the Graduate School and University Center, both part of the City University of New York, have provided indispensable support—intellectual, moral, and financial—over a great many years. I am grateful also to Wolfgang Hirschberg, who taught me how to read Russian. For their general encouragement and guidance of my research, I would like to thank Joseph W. Dauben, Evelyn Ackerman, Philip Dawson, David Nasaw, James R. Jacob, Margaret L. King,

[13] Joan Debardeleben, *The Environment and Marxism-Leninism* (Boulder, CO: Westview Press, 1985); Marshall Goldman, *The Spoils of Progress: Environmental Pollution in the Soviet Union* (Cambridge, MA: MIT Press, 1972); Philip R. Pryde, *Conservation in the Soviet Union* (Cambridge: Cambridge University Press, 1972); and idem., *Environmental Management in the Soviet Union* (Cambridge: Cambridge University Press, 1991); John Stewart Massey, Ed., *The Soviet Environment: Problems, Policies and Politics* (Cambridge: Cambridge University Press, 1991).

[14] I. Bergkholtsas, *Baltika: Okhrana prirodnoi sredy i pravo* (Riga: "Avots," 1986); M. Ia. Lemeshev, *Bureaucrats in Power: Ecological Collapse.* Trans. Sergei Chulaki (Moscow: Progress Publishers, 1990).

Bernard Semmel, and Joel Weiner. For getting me interested in this topic in the first place, and for his comments on some of my research on it, I would also like to thank Douglas R. Weiner of the University of Arizona at Tuscon. I am further indebted to Marc Raeff and Allen McConnell for close reading and useful comments on an earlier version of this book. Thanks and love to my wife, Lena, who for more than sixteen years has put up with the manifold consequences of my decision to pursue an academic life, and has supported me unfailingly throughout, and to our daughter, Isabel.

I save for last a very special debt of gratitude: to my mentor and friend Abraham Ascher. You have been my best and most patient teacher. I thank you from the bottom of my heart.

Research for this book was supported in part by a grant from the International Research and Exchanges Board (IREX), with funds provided by the National Endowment for the Humanities, the US Department of State, and the US Information Agency. None of these organizations is responsible for the views expressed. Funding was also provided by the Department of History and the Provost's Office at the Graduate School and University Center of the City University of New York, and by the University Research Council at the University of Central Arkansas. To these offices also I am most grateful.

Note on dates and style

On 31 January, 1918, Russia adopted the Gregorian or Western calendar (replacing the Julian version in use since the time of Peter the Great). Dates given in this book are those officially obtaining in the country at the time and are indicated, where relevant as New (Gregorian) or Old (Julian) Style.

This book uses the Library of Congress transliteration system, though the usual exceptions of Anglicization have been made for well-known names and terms (Trotsky, not Trotskii, for example).

II
Forest Conservation in Imperial Russia

The forest is part of the very fabric of Russian life, both material and cultural. Pre-Kievan Slavic communities and countless generations of Russians after them sought its sanctuary in the face of periodic nomadic invasions of the open steppe. Then and since, much of the day-to-day business of building and defining Russia has gone on in the forest. The great Russian historian V. O. Kliuchevsky was far from alone in attributing to the "swamps and forests" the "racial character of the Great Russian"—born, in his phrase, of the settler's confronting therein "a thousand unforeseen risks, difficulties and hardships," all of which taught him to "watch nature very closely."[1] In these forest fastnesses elaborate systems of belief long ago evolved among Slavic communities based on reverence for the forests as a supernatural environment inhabited by a variety of spiritual entities. The residue of these beliefs permeates Russian folk and literary culture even to the present, imbuing this natural resource with a lingering mystique.[2]

Later, as Russians looked outward rather than in, the tree-clothed eastern horizon became a goad to expansion. During Europe's era of discovery and conquest, as men like Columbus and da Gama plied the great waters of the Atlantic and Indian oceans, their Russian counterparts, utilizing rivers and portage, sailed east across the sylvan expanses of Siberia, colonizing as they went. Iberian and Russian alike were primed for expansion through conflict with Islamic opponents (the Moors in the case of the Iberians, the Tartars for the Russians). Each was lured by the possibility of lucrative trade with Eastern markets. And each, also, discovered along the way other opportunities, which they dallied to exploit with abandon. The Spanish arrived in America, where they

---

[1] V. O. Kluchevsky, *A History of Russia*. Trans. C. G. Hogarth. Vol. I (New York: Russell and Russell, 1960), 218.

[2] For a useful and brief overview of traditional Russian (and Siberian) folk beliefs associated with the forest see chapter one of Charles E. Ziegler, *Environmental Policy in the USSR*. (Amherst: University of Massachusetts Press, 1987). For more detailed treatments see Marjorie Mandelstam Balzer, *Russian Traditional Culture: Religion, Gender, and Customary Law* (Armonk, NY: M.E. Sharpe, 1992).

found gold and silver to plunder. The Russians discovered forests seemingly without end—and within them untold riches of furs and skins. It was during, and largely because of these discoveries that Russia emerged as a world power—a fact further cementing the symbiosis of forest, people, and state.

Russia is a forest nation also by simple geography. Current figures from the European Forest Institute show precisely 50% of the territory currently within that country's borders as "forest," with a further 4% designated "other wooded land." This total translates to over 886 million hectares.[3] Border changes, the greater inaccuracy of earlier methods of measurement, varying definitions of the term "forest," and even the question of whether one is counting Russia alone or the Russian Empire all problematize direct comparison of even the most basic current and historical Russian forest statistics. Contemporary estimates for the 1920s, where this study is largely focused, established Russia's forest area in a range from about 660-770 million hectares—noticeably smaller figures.[4] At the same time, and still more confusingly, figures given in the 1920s put the *percentage* of Russian territory under forest as high as 68%.[5] Presumably, a very generous definition of the word "forest" was used in this reckoning. Estimates from the 1950s and 1960s typically come in somewhere between the above extremes. The 3rd edition of *The Great Soviet Encyclopedia*, for example, reckoned the Russian Republic's total forest area around 1960 at 705.9 million hectares[6]). Whichever statistics are assumed, Russia in the 1920s could certainly claim approximately one-fifth of the world's total forest cover.[7] With significant variations over time, this patrimony was owned or administrated in a wide variety of ways. During the eighteenth, nineteenth, and pre-Revolutionary twentieth

---

[3] The European Forest Institute, "Forest Resources of Russia." http://www.efi.fi/fine/resources/russia.html (21 March, 2004).

[4] By way of example, the figure of 611,917,000 *desiatinas* is given for 1924 in a Russian trade journal. Since 1.0 *desiatina* equals 1.0925 hectares, this is about 668 ½ million hectares. RSFSR. *Vserossiiskii tsentral'nyi ispolnitel'nyi komitet*. XI sozyv. Tret'ia sessiia. Zasedanie vos'moe (15 October, 1924), 348.

[5] *Lesnoe Khoziaistvo SSSR za 50 let, 1917-1967*. Gosudarstvennyi komitet lesnogo khoziaistva soveta ministrov SSSR (Moscow: Lesnaia promyshlennost', 1967), 51.

[6] Compare to a figure of 820.4 million hectares (broken down as 180.2 million in European Russia, 8.2 million in the Caucasus, and 632 million in "Asian Russia"), published in ibid., 31.

[7] RSFSR. *Vserossiiskii tsentral'nyi ispolnitel'nyi komitet*. XIV sozyv. *Biulletin'* 8. Zasedanie vos'moe (24 November, 1929), 5.

centuries with which this chapter will be concerned, the major players were the state, the crown, private individuals and entities, and peasant communes.

Immediately prior to the Bolshevik Revolution itself, timber may have accounted for as much as a third of all Russian exports measured in rubles earned.[8] A 1916 official report estimated total timber exports for 1913 of 461 million *pudy*[9] (about 7.6 million metric tonnes). During the previous several years the figure had been only slightly lower (see table 1).

Table 1: Russian Empire Timber Exports, 1909-1913 [10]

Year	Quantity (Millions of *pudy*[11])	Value (Millions of Rubles)
1909	423	126
1910	416	138
1911	414	142
1912	425	152
1913	461	164
1914	224	103

[8] RSFSR. *Vserossiiskii tsentral'nyi ispolnitel'nyi komitet*. XI sozyv. Tret'ia sessiia. Zasedanie vos'moe (15 October, 1924), 363.

[9] *Lesopromyshlennoe delo* 5-6 (1922): 1. These figures should not be regarded as definitive, however. Subsequent (Soviet-era) estimates of late-Imperial economic indicators for the forest sector are often lower—whether correctly or for political purposes is difficult to say. For example, a 1928 Soviet trade journal revised 1913's total timber exports down from over 460 to only 350 million *pudy*. See, "O deistvitel'nykh razmerakh russkogo dovoennogo lesnogo eksporta," *Lesnoe khoziaistvo i lesnaia promyshlennost'*, 8-9 (1928): 49. A *pud* weighs about 36 pounds.

[10] RSFSR, Ministerstvo finansov, *Narodnoe khoziaistvo v 1914 g* (Petrograd, 1916), 217.

[11] One *pud* equals about 36 pounds.

The well-being of Russia's peasant population, too, traditionally rested heavily on the forests. Up to and even beyond the Bolshevik Revolution peasants, "built with pine and oak . . . heated with birch and aspen . . . shod themselves with bast and . . . made household tools of linden."[12] Even the typical peasant plate, spoon, and cup were made of wood. As late as 1924, peasants remained the single greatest consumers of timber, accounting for half of all demand.[13] An anonymous writer in 1917 for the journal *Lesnaia zhizn' i khoziaistvo* (Forest Life and Economy), viewed the average Russian's traditional dependence on the forest in the following terms:

> Some [go into the forest] to get firewood, or building materials, others to feast on berries or collect mushrooms. Many [visit] to earn a little money, since for working people the forest . . . is always a place to find work: cutting timber, transporting it, or something else. . . . [The forest] gives those who are not wealthy the opportunity to have their own fuel and safety from the cold.[14]

At the same time, thousands upon thousands of domestic furnaces, stoves, and boilers were powered almost entirely at the expense of the forest—as, moreover, were most of the trains and many of the mills, plants, and factories of Russia's early, sporadic industrialization.

Since the population—and thus all this forest activity—was not spread evenly across the country but crowded, for the most part, into a core area of European Russia, it is understandable that the growing manifold demands for wood sooner or later would begin visibly and seriously to affect large portions of the forests themselves, and equally understandable that the idea of forest conservation would make its eventual debut. In the European order of things, Russia came to such a moment relatively late—generally speaking, in the second half of the nineteenth century (though with some important if unsustained eighteenth-century antecedents, discussed below). Nations such as Britain, France, and Germany—which had cleared their smaller original forests earlier and more completely—had already begun to develop ideas, traditions, and programs of forest management, regeneration, and conservation from the seventeenth century, even if they too became much more serious about them only

---

[12] V. O. Kliuchevskii, *The Course of Russian History*. Quoted in Teplyakov, et al., *A History of Russian Forestry and its Leaders*, v.

[13] V. Solov'ev, "Krest'ianstvo i lesnye rabotniki," *Lesovod* 4-5 (1924): 9.

[14] "Beseda o lese i ego znachenie v zhizni naroda," *Lesnaia zhizn' i khoziaistvo* 2 (1917): 5.

in the nineteenth century.[15] One sign of a growing interest in forest conservation was that not only foresters or other directly interested parties took up the cause. The great Russian playwright Anton Chekhov advocated for forest conservation through his literary characters. *Uncle Vanya's* Astrov, for example, asserted that:

> forests glorify our earth, they teach a person the meaning of beauty and kindle a spirit of majesty. The forests make the climate less severe. In countries where the climate is mild less energy is spent in the struggle with nature, and so there we find a gentler, more understanding person. There too, people are beautiful, pliant, and responsive; their speech is elegant, and their movements are graceful. Both science and art thrive among them, their philosophy is never somber, their attitude toward women is fully refined, noble . . . .[16]

In the play, this romantic outburst is dismissed by the character Voynitsky, standing in, in this regard, for Russia's peasants, entrepreneurs, and other parties not interested in conservation: "Bravo, bravo! . . All of this is very nice but it's not convincing. So . . . let me go on burning firewood in my stoves and building my barns out of trees."[17] The dynamic of this short conversation—high-minded and socially-principled moral exhortation summarily swatted by practical self-interest—reflects not only the individual personalities of Chekhov's characters, nor even merely a wider late nineteenth-century reality, but (as will be seen) aptly describes much of the overall interaction of forestry apparatus and rural population throughout the early Soviet period, too. Nor was Voynitskyian complacency the extreme of contemporary anti-conservationist sentiment, even among intellectuals. Several times during the pre-Revolutionary twentieth century it was argued that vast forests and abundant, cheap timber was in fact Russia's curse, not its blessing—that it discouraged the transition to coal and oil fuel, to cement and brick construction, and in general to modernity and prosperity.[18]

---

[15] The French forest code of 1669 stands out particularly in this regard. See chapter 10, n. 8.

[16] Anton Chekhov, *Uncle Vanya*. In *Anton Chekhov's Plays, A Norton Critical Edition*. Trans. and Ed. Eugene K. Bristow (New York: W. W. Norton and Co., Inc., 1977), 63-64. The quoted section is actually spoken by another character, Sonya, in summary of Astrov's ideas as she remembers them.

[17] Ibid., 64.

[18] See for example, V. F. Kliuchnikov, "Konkretnye zadachi lesookhraneniia," *Lesnoi zhurnal* 9-10 (1911), 1414-15; K. Danishevskii, "Novyi kurs," *Biulleten' tsentral'nogo upravleniia lesnoi promyshlennosti (glavleskoma) i tsentral'nogo komtiteta profsoiuza derevoobdelchikov* 10-11 (1922): 1;

More, not less cutting was hoped for here.

Nonetheless, forest conservation impulses did develop in Russia. In approaching their early history it is useful to begin about the turn of the eighteenth century. Though a few legislative or other antecedents date back as far as Kievan times, these were essentially assertions of royal patrimony, designed not to protect the forests *per se*, but to assert the power and property rights of the monarch therein.[19] These were measures generally bereft of any real consciousness of forests as scarce—and thus precious—resources, a fundamental prerequisite of any conservation worthy of the name.[20] Thus defined, forest conservation's origins lie, as does so much else in Russian history, with the remarkable Peter the Great.

Peter the Great and the Beginnings of Forest Conservation

In his quest to make Russia a great power, and to this end to give it a powerful navy, Peter began in 1701 to protect some of its best sources of ship-timber—a prudent move, perhaps, given that a single battleship of the time required up to three-thousand trunks.[21] A decree of that year forbade clearing forests anywhere within thirty *versts*[22] of specified rivers. More decrees followed in 1703. One issued in April established the *Izmailovskii* forest reserve (*lesnoi zapovednik*) near Moscow. One in November decreed the protection—at the

---

and A. S. Ermolov, *Sovremennaia pozharnaia epidemiia v Rossii* (1910).

[19] A quick overview of Pre-Petrine forest law is given in V. K. Teplyakov, et al., *A History of Russian Forestry and its Leaders*, chapter 1. A more in-depth treatment (in Russian) can be found in N. I, Faleev, "Starye lesnye osnovy," *Lesa respubliki* 5-7 (1918).

[20] Useful overviews of pre-Petrine forest legislation are given in V. E. Boreiko, "Plakala Sasha kak les vyrubili," 6-8; and in N. I. Faleev, "Starye lesnye osnovy," *Lesa respubliki* 5-7 (1 June, 1918), 265-270. A much more detailed treatment of Kievan, Muscovite, and Imperial Russian forest law, with an emphasis on ownership and jurisdiction is given in I. Gershman, "Ocherk istorii lesovladeniia, lesnoi sobstvennosti i lesnoi politiki v Rossii," *Lesnoi zhurnal* 3-4 (1911). Also of use is N. I. Faleev, ed., *Osnovnoi zakon o lesakh, izdanie neofitsial'noe* (Moscow: Izdanie Narkomzema, 1918).

[21] V. E. Boreiko, "Plakala Sasha kak les vyrubili," 8.

[22] A *verst* is equal to 1.067 km or about 3500 feet.

threat of execution—of oak forests near Simbirsk and Kharkov.[23] This latter decree also identified and called for inventorying of forests—typically along rivers—whose trees, particularly oak, elm, ash, larch, and pine were sufficiently large (44 centimeters trunk width) to provide good sources of naval timber.[24] Several other similar decrees were issued over the following decade and for like purposes. In 1712 Peter banned timber cutting "within ten *verst*s of the banks of the rivers in the Novgorod, Starorusskii, Lutskii, and Goronetskii *uezd*s."[25]

In 1722 Peter issued a much more comprehensive set of instructions governing forest cutting and protection. In certain aspects these provided a general basis for subsequent Russian forest legislation at least up to the Bolshevik Revolution. Particularly influential was the Instructions' creation of two major forest categories, "protected" and "unprotected (*zapovednye* and *nezapovednye*).[26] A sub-category of protected forests (which in later tsarist legislation came to constitute a clear third category) was also designated. Entitled *vodookhranye* ("water-preserving") forests, these were areas deemed of significance in

---

[23] V. K. Teplyakov, et al, A History of Russian Forestry and its Leaders, 3.

[24] V. N. Makarov, *Okhrana Prirody v SSSR*. (Moscow: Gosudarstvennoe izdatel'stvo kulturno-prosvetitel'noi literatury, 1947), 55.

[25] V. E. Boreiko, "Plakala Sasha kak les vyrubili," 9.

A *uezd* (sometimes translated as "district") was a jurisdictional unit in pre- and post-Revolutionary Russia. In the former setting, as here, it was a sub-division of the larger *guberniia* (province), and was itself further divided into *volost*s. The largest territorial units were *oblast*s, or regions. Thus, the following hierarchy of subdivisions existed: *oblast*s were divided into *guberniia*, these into *uezd*s, and these into *volost*s. The same holds true for the early Soviet period also. Since some of these terms translate into English better than others I have elected throughout this study to retain the Russian terms *uezd* and *volost*, and to use the English equivalents for the larger units.

[26] V. N. Makarov, *Okhrana Prirody v SSSR*, 19. The Russian language contains several different terms, all of which are usually rendered in English as "protected." Thus *zapovednye* and *nezapovednye* would be translated as "protected" and "not protected." Later, but pre-Soviet forest legislation used the word *zashchitnyi* (which also means protected). The word *zapovednik*, used throughout Russian and Soviet history and usually translated as "nature preserve," will also be encountered. *Zapavednyi* and *zapovednik* derive from the word *zapoved'*, which literally means "commandment" or "testament," and thus may have a kind of other-worldly or religious overtone absent in the more neutral term *zashchitnyi*.

protecting waterways and water-sources, by minimizing soil erosion. Under Peter, *vodookhranye* forests were declared along riverbanks. Depending on the size of the river itself, the protected strips ran from 15-55 *versts* in width. In them, cutting was banned completely. The three forest categories thus established—"protected," *vodookhranye*, and "all others"—remained (with varying fortunes) in existence up to 1917. Perhaps Peter's last forest protection decree was issued in 1723. Also intended to conserve naval supplies, it mandated a temporary total ban on the cutting of certain forests designated as "ship" timber *(korabel'nyi les)*.

Beyond the legislative realm, Peter acted in other ways to protect forest resources. The historian V. E. Boreiko, for example, relates anecdotally that on one occasion, "in order to provide to the public an example of his respect for oak forests, [Peter] . . . fenced off two mature oak trees he had come across in Kronstadt, [then] set up a round table and some seating under them."[27] Similarly, Peter also planted trees with his own hands, supposedly to encourage others to do the same.[28]

The formidable Peter seems also to have been responsible for establishing in Russia the practice of cutting timber according to what was generally known in western Europe at the time as *La Methode à Tire et Aire.* This technique, pioneered in France and on which there were several variations, involved estimating the number of years a given area of forest needed to reach a harvestable age and then dividing it into that number of sub-zones. These sub-zones would then be cut one per year so as to ensure that the time taken to cut them all equalled the time taken for full regeneration of the first zone, which would then be ready for cutting again. This theoretically ensured an indefinite supply of mature timber. Used widely in western Europe during Peter's time, *La Methode à Tire et Aire* was incorporated into his decree of 1722 which required that forests assigned for the use of specific plants or mills be divided into between twenty-five and thirty "cutting areas" [*lesoseki*], only one of which could be cut in any one year.

Peter's approach to forest protection was, of course, practical rather than sentimental (which is also true of Lenin, two hundred years later, see chapter 8). He saw forests as sources of timber: primarily for shipbuilding, but also for iron smelting and for export. Accordingly, he gave jurisdiction over forest matters to the Admiral Collegium, a naval body. He was motivated overwhelmingly by the

---

[27] V. E. Boreiko, "Plakala Sasha kak les vyrubili," 10.

[28] Ibid.

needs of state and wanted to prevent these being compromised by timber shortages. Violation of forest law, accordingly, was viewed as activity directed against the state, not against the environment, and as such resulted in the harsh punishments noted. Significantly, when the needs of state demanded cutting rather than conservation, the Russian emperor had no thought for sparing trees, which he always understood ultimately as means to other ends. To aid troop movements and for other strategic purposes during his battles with Sweden's king Charles XII, for example, Peter did not hesitate to order major forest clearings in 1708.[29] At the same time, however, he is reputed generally to have had also at least some proto-ecological smarts, suggested best, perhaps, in his designation of *vodookhranye* forest zones as a measure against soil erosion. Here too, however, Russia's political and military power was his chief concern: unnavigable waterways hindered ship and log movements, stifled trade, and weakened the state.

Post-Petrine Forest Legislation to 1888

Not for over one-hundred-and-fifty years after Peter's death was Russia to see forest legislation of comparable significance, though a number of minor developments are worthy at least of passing note, many during the reign of Catherine the Great (1762-1796).[30] In 1766, for example, a general survey of Russian forests was announced, leading to the creation of the first reliable maps and thus opening up new possibilities for management and conservation.[31] And in 1774 a decree banned forest burning in Siberia in areas known to be heavily inhabited by sable. This was, of course, aimed at protecting the fur trade rather than the forests, though it may well have benefited the latter as well.

Greatly more important during the second half of the eighteenth century was the emergence of what would in the late nineteenth and early twentieth centuries be derided by foresters and others as "commercial" (*kupecheskyi*)

---

[29] Ibid., 11.

[30] A careful examination of Russian legislation during the eighteenth and nineteenth centuries reveals dozens of decrees directly related to forest protection and literally hundreds having at least some connection. Obviously only a select few can be treated here.

[31] V. K. Teplyakov, et al., *A History of Russian Forestry and its Leaders*, 4.

forestry, especially the growing practice whereby forest land would be purchased by an individual or concern looking to profit by felling the trees, selling the timber, and turning the land to agriculture. In this manner a general situation arose during the eighteenth century, and persisted more or less unchanged until the late nineteenth century, whereby public forests received some level of protection and rational management while private forests were open to essentially unregulated exploitation. These developments derived in large part from a series of decrees issued between 1769 and 1801 by Catherine the Great, Paul I, and Alexander I providing increasingly secure private property rights—including in forest land.[32] A law of 1769, for example, allowed the sale of privately-owned forest land after first dividing it into thirty plots. The Decree of 22 September, 1782 gave owners of private forest land "complete freedom to use all forests to their greatest advantage."[33] By releasing the nobility from various obligations to the state the new decrees also undermined many of the Petrine protections already outlined. In particular, rights were restored freely to exploit privately owned forests previously protected for the shipbuilding industry. Writing in 1925, the forest scientist P. Morozov linked to this period, and to Catherine in particular, "a century-and-a-half of the most absurd, reckless abuse of so-called private forests."[34]

Not only private owners but the state also began at the close of the eighteenth century to look more closely at the forests as sources of revenue and profit. Forest-use taxes were introduced in 1799, establishing thereby a direct link between exploitation and state revenue.[35] This was followed in 1800 by a division of all Russia's forests into eight basic economic and natural zones in anticipation of more organized and efficient use. The new more exploitative outlook suffused the 1802 "Charter on Forests" (*Ustav o lesakh*)—a comprehensive forest code partly worked up under Catherine the Great but which had stalled until passed now under the administration of Alexander I. The Charter's major policy goals were to "make forests profitable for the state" and to end timber shortages.

---

[32] On the rise of private forest property in the Russian Empire in general, including the effect of these decrees, see: N. I. Faleev, "Kapitalisticheskye osnovy lesnogo khoziaistva." In N. I. Faleev, ed., *Osnovnoi zakon o lesakh, izdanie neofitsial'noe* (Moscow: Izdanie Narkomzema, 1918), 12-13.

[33] V. K. Teplyakov, et al., *A History of Russian Forestry and its Leaders*.

[34] P. Morozov, "Chto delat'?" *Lesnoi khoziaistvo, lesopromyshlennost' i toplivo,* 9-10 (1925): 5.

[35] *Lesnoe Khoziaistvo SSSR za 50 let, 1917-1967*, 234.

Notably, however, efficient use and general protection were also prioritised.[36] The legislation was accompanied by the creation of Russia's first fully-dedicated forest organ, the Forest Department. In accordance with the new code's priorities, however, and in a clear break with Petrine tradition, the Department was established not under naval control—as previous forest administration had been—but was put instead under the Ministry of Finance. (A subsequent transfer of the Forest Department in 1811 put it under the control of the Department of State Domains, later upgraded to a Ministry, where it remained, with some reforms, up to 1917.)

A modest move in defence of the forests came in 1832 with the issuance of a "Statute on a Permanent Forest Conservation Service." Intended primarily to recruit and retain forest conservation workers, the law offered a twenty-year draft in exchange for exemption from state taxes and military and civil service requirements.[37] In the same direction, over the coming decades a growing emphasis was placed within official forestry circles on the need for "rational forest management." The most important statement of this new priority came in an Imperial Order to the Ministry of State Domains (*Ministerstvo gosudarstvennykh imushchestv*) in 1855.

While these and other similar developments in mid-century resulted in modest increases in forest personnel, amounts of land reafforested, and successful prosecutions for forest law violations, they were swamped by the rising tide of forest cutting and degradation overall, especially following the peasant emancipation of 1861. The abolition of serfdom—which largely failed in its intended goals of creating a more modern, patriotic, and educated population, and thus also in undergirding the creation of a more powerful and prosperous Russia following the state's humiliating defeat at British and French hands in the Crimean War—was also a negative development for forest conservation. Much of Russia's existing apparatus for carrying out basic forest work—cutting, planting, tackling fires, and so on—had rested on serf obligations. Free peasants could no longer be counted on to carry out these and many other relevant responsibilities

---

[36] The *Ustav o lesakh* concerned forest administration protection and planting. See, *Polnoe sobranie zakonov Rossiiskoi imperii* (1802), 350-55. Alexander also passed a law "On the preservation of protected forests from cutting" (*Ob okhranenii ot porubki zapovednykh lesov*) some months earlier the same year. See ibid, 254-57.

[37] V. K. Teplyakov, et al., *A History of Russian Forestry and its Leaders*, 6.

and duties established in the days of serfdom. Efforts made beginning in 1869 to make up the deficiency by recruiting paid forest guards had little apparent effect. Upheavals associated with land redistribution, peasant movements, and so on, also tended to result in an increase in unauthorized cutting. At the same time, by depriving landlords of their traditional serf laborers, the Emancipation also created a wave of landlord impoverishment. Hard-up landlords often cleared land in order to sell the timber and gain more land for agriculture. Others sold their estates to purchasers who did the same.

Though only about one-third of European Russian forest (and less in Siberia) came under private ownership up to the First World War (see table 2), the growth of private proprietorship as a general trend has been interpreted as having engendered a "total revolution in the minds of [forest]-owners," allowing and encouraging the trade—both within and beyond Russia—of all sorts of forests, including officially protected ones,[38] and thus as further undermining existing levels of conservation deriving from the Petrine period. Over the course of the nineteenth century, and to a degree perhaps unparalleled in the west, private forestry came to be reviled by influential members of the major forest societies—and even within the Forest Department itself—as a continuing and primary threat to good forest management and to the existence of the forests themselves, helping create early, scattered momentum toward some kind of "socialist" alternative to the "excesses" of private forestry (see further below). Doctor Shingarev—noted expert and author on peasant affairs and later to become agricultural minister of the Provisional Government—wrote in 1907, for example, of "intrigues [within the *Duma*[39]] of those interested in the maintenance of private property in land." He also quoted a Committee report of the same year on over-cropping, which cited deforestation as a major threat to Russia's agricultural and general wealth: "Since the Emancipation, the woods have been cut down (owing to the extension of cultivation), the streams have become shallow or have disappeared, drifting sands have invaded the fields . . . and instead of once fertile lands, gullies, water-courses, landslips and even precipices, have made their appearance: the land has lost its fertility, its productiveness has fallen, the natural wealth is exhausted, and the people are impoverished."[40] This seems to have been a common opinion of the time.

[38] N. I. Faleev, "Starye lesnye osnovy," 268. The opinion expressed by Faleev was common throughout much of the late nineteenth and early twentieth century.

[39] Russian Parliament, established as a major consequence of the Revolution of 1905.

[40] Sir John Maynard, *The Russian Peasant and Other Studies* (New York: Collier, 1942), 68-69.

Table 2: Landholdings in 1913 (Russian Empire) [41]

Government	48.8 %
Royal Family	5.0 %
Private	31.9 %
Peasant Communes (*Obshchestva*)	11.6 %
Other	2.7 %
Total	100 %

Anxiety about the growing pace of cutting seems not to have been unfounded. Late Imperial and Soviet sources are here in general agreement about the pace of cutting during the late nineteenth and early twentieth centuries, with a typical estimate being that between 1867 and 1914 as much as thirty percent of the entire forest area of European Russia was cut down[42]—perhaps the worst of these excesses occurring in the earlier part of this period. During the mid-nineteenth century, also, forests that had remained relatively untouched for centuries began to fall under increasing industrial exploitation. Official permission to cut in Karelia, for example, was given for the first time in 1863.[43]

It was largely in response to this accelerating deforestation (and in the estimation of Revolutionary-era foresters such as N. I. Faleev—a specialist on forest law at the Petrograd Forest Institute and future head of the Bolshevik forest department[44]—due also to a growing awareness by the government of forests'

[41] N. Knipovich, *Ocherk deiatel'nosti NKZ-a za 3 goda (1917-1920)* (Gos-izdatel'stvo, 1920), 20.

[42] *Lesnoe Khoziaistvo SSSR za 50 let, 1917-1967*, 4.

[43] L. G. Chertov, "Ekonomicheskaia otsenka lesov Evropeiskogo severa SSSR i prirodnykh uslovii ikh eksploatatsii." In *Geograficheskii sbornik XI: Ekonomicheskaia geografiia* (Moscow and Leningrad: Izdatel'stvo Akademii nauk SSSR, 1957), 26.

[44] Faleev was also the chief author of the Bolsheviks' first major piece of forest legislation, 1918's "On Forests." He will feature prominently throughout much of this study.

various climatic, economic, and social functions[45]) that impetus began to build for the issuance of new and better forest protection codes. In this connection, Alexander II called upon the Ministry of State Domains in late 1864, though with no apparent result. Only in 1882 was a modest measure, "On Strengthening Accountability for Unauthorized Cutting" (*Ob usilenii otvetsvennosti za samovol'nuiu rubky*) passed by his successor Alexander III. This established a scale of fines for violators, increasing with each successive conviction from twice the value of the damage done to three times plus a six-month prison term. A comprehensive approach to better forest protection, under the title "Statute on the Safeguarding of Forests" (*Polozhenie o sberezhenii lesov*[46]) emerged only in 1888. It marks, however, a watershed in Imperial Russian forest protection legislation and thus deserves special attention.

Coming to Grips with Growing Forest Destruction: The "Statute on the Safeguarding of Forests," April 4 1888 [47]

With certain amendments, expansions and special instructions, the "Statute on the Safeguarding of Forests" remained the basis, and the most powerful statement, of forest protection in the Russian Empire until the First World War.[48] Covering large parts of European Russia, but not the more richly-forested areas of the far north or Siberia,[49] the Statute aimed to establish measures "guarding . . .

---

[45] N. I. Faleev, "Starye lesnye osnovy," 268.

[46] The word "sberezhenie" presents a translation problem. It is typically rendered as "conservation," as is the Russian word "okhrana" (*okhrana prirody* is the standard term for what is in the west understood as "nature conservation." To distinguish between two words both of which can be identically rendered in English, I have chosen to translated *sberezhenie* as "safeguarding," which more accurately connotes the meaning of the word as employed in this decree.

[47] "Polozhenie o sberezhenii lesov," *Sobranie uzakonenii i rasporiazhenii rossiiskogo pravitel'stva* (1888), 406, 831-46.

[48] The 1888 law was promulgated with numerous geographical exemptions even within the limited sphere of European Russia. These exemptions were somewhat reduced over the following years as the law was introduced in more areas. By 1914, however, many forests remained beyond its jurisdiction, including most importantly those in Siberia, and most parts of the Archangel region, and elsewhere.

[49] As originally passed, the decree covered forests in European Russia and the Caucasus totaling an

against [forest] destruction and exhaustion," to encourage their "correct economic management," and to facilitate the raising of new forests (articles 1-2). To this end it revamped the Petrine idea of establishing "protected" forests,[50] which it defined as those of vital interest to state or society and in need of "special safeguarding measures" (article 3). Significantly, these could henceforth be declared on any type of land at all regardless of its ownership "by the state, crown, various establishments, [peasant] communes, [or] private individuals" (article 1). The wide reach thus given the law, at least in the geographical areas in which it operated, was one of its most positive attributes.

Protected forest areas could not legally be cleared or turned to any other type of use. Following designation, a ban obtained also on the cutting of any and all "growing trees." This high level of protection was to be temporary, however, pending the creation in each specific case of an "economic plan," to be worked out by local Forest Preservation Committees (created for this purpose and by this legislation)[51] in cooperation with the relevant forest owner and in accordance with general principles subsequently to be laid down by the Ministry of State Domains (articles 6, 14). Economic plans, in turn, were required to promote a system of "gradual" cutting facilitating "natural" and continual regeneration (article 7).

This early example of forest planning, which sought to combine local interests and on-the-spot evaluations with general central guidelines, fell short of

---

area of some 34 million *desiatinas* (about 37 million hectares). As exclusions from the operation of the law within European Russia were reduced (it was introduced into Poland, for example, in three stages from 1899 to 1901) the total rose to some 55 million *desiatinas* (slightly over 60 million hectares) by 1914.

[50] The term "zapovednyi," which had been used in legislation during Peter the Great's era was here abandoned in favor of the more secular and legalistic-sounding "zashchitnye." Both are typically translated as "protected," however.

[51] Constituted at the provincial and regional (*gubernskyi* and *oblastnoi*) levels (articles 23-24), these were to consist of a variety of officials including representatives of the local court (*okruzhnoi sud*), Ministry of State Domains, royal landholdings, and provincial council (*gubernskaia uprava*), plus two local forest owners to be chosen by a provincial land assembly (*gubernskoe zemskoe sobranie*, see article 25). Forest Protection Committees convened specifically on matters relating to the protection of river sources and headwaters (*verkhoviia*) were to include also a representative of the Department of Transport (*Vedomstvo putei soobshcheniia*).

its promise in several ways. For example, the words "may be" [*mozhet byt'*] crop up frequently enough, and in suitable location throughout key passages of the decree so as to create legal loopholes in fact appearing in some places to allow the clear-cutting expressly forbidden elsewhere. What actually was allowed *in situ*, it seems, depended mostly on the inclinations and abilities of local authorities, which varied considerably. Economic plans also forbade (though with similar ambiguity) other destructive practices, including the grazing of cattle on young shoots, the collection of straw and bedding (*podstilki*) and other secondary forest usage. In addition to declaring a forest area officially protected as noted, Forest Preservation Committees also had the option—where preservation was deemed necessary to protect headwaters and river sources—to establish an area as "subject to safeguarding" (article 29). These forests were elsewhere and subsequently referred to, as in Petrine legislation, as *vodookhrannye*, "water-preserving," forests.[52]

Forests not declared protected or water-preserving—the vast majority—were nonetheless also accorded legal protections by the Statute, though of dubious worth. For example, while the "clearing of forests" was banned universally in principle, very broad exemptions were given, including "when [clearing] is necessary in order to establish the more profitable organization of an estate" (article 11). On the other hand, it was required also that in all cases permission for clearing be sought in advance from the local Forest Preservation Committee; and a six- to twelve-month waiting period was mandated between the original request and issuance of a permit. In the case of privately-owned forests, requests could still be denied, estate profitability notwithstanding, if the Forest Committee deemed that the potential destruction to the forest economy of the wider region would be unacceptable (articles 11-13). A few other useful restrictions obtained also.[53]

---

[52] "Water-preservatory forests" seem to have differed from the protected forests already noted only very slightly, essentially in that where the former were areas to be exploited on a sustainable basis according to an economic plan, water-preservatory forests, were intended to be left standing so as to hold soil together and thus prevent erosion. Cutting permits could, however, be arranged in these forests too, along the same lines as obtained regarding protected forests (articles 29-31, passim).

[53] Responsibility for overseeing the fulfilment of the various forest preservation measures so far outlined lay ultimately upon the Forest Department within the Ministry of State Domains *(Ministerstvo gosudarstvennykh imushchestv).* Actual oversight in the field, the designation of forest areas as protected, and evaluation and policing of forest activities overall was made the responsibility primarily of the local

In general, as a direct result of this legislation, where unauthorized cutting could be proved to have occurred—in forests of *any* designation—violators were legally required at their own expense to reafforest the affected areas within a timeframe to be established by the Forest Preservation Committees. Failure to comply resulted in completion of the work by the local Forest Preservation Committee itself at the owner's expense (article 15).[54] More significantly—both in terms of facilitating protection and as an indication of things to come, in instances where private or communal proprietors refused or were unwilling either to carry out or to pay for "necessary" conservation measures, the decree authorized the compulsory sale of their land to the state (article 9).[55] Such purchases, moreover, seem on numerous occasions over the following decades to have been made. More importantly, against a background of growing condemnation by foresters of private forest practices, the notion that the state should take possession of sensitive or abused forest land for the good of the wider society enjoyed growing popularity and became a constant theme in forestry journals and congresses in the early twentieth century. (It was not, of course, so

Forest Preservation Committees.

[54] As further conservation measures, the "Statute on the Safeguarding of Forests" also envisioned the selection and deployment, albeit here at the invitation of private owners, of "experienced persons" to give "technical assistance" for the purpose of encouraging better forestry practices. It noted also that foresters and forest guards were to be given awards and recognition for outstanding or long service (articles 20-21.) These awards—usually in the form of gold or silver medals—had obtained earlier, but were increased in 1888, and again subsequently. According to Boreiko, thirty were given out between 1881 and 1894 (V. E. Boreiko, *Belye piatna istorii prirodookhrany SSSR, Rossia, Ukraina, tom vtoroi*. Seriia: Istoria okhrana prirody, vypusk 7 [Kiev: Kievskii ekologo-kul'turnyi tsentr, Tsentr okhrany dikoi prirody SoES, 1996], 25). Further legislation passed over the next few years, building on the "Statute," provided for increased numbers of forest guards, often in certain very specific areas, such in the vicinity of a single timber plant or other factory. See for example, "Ob usilenii strazhi dlia okhrany lesa, prilegaiushchego k Kazanskomu porokhovomu zavodu," *Sobranie uzakonenii i rasporaizhenii rossiiskogo pravitel'stva* (1892), 281: 505-06.

[55] Owners dispossessed in this fashion could repurchase their lands at any time up to ten years after the initial loss for their original sale price (with 6% per year interest), plus the cost of the conservation measures carried out by the state (article 9).

popular with forest owners themselves.) Overall, this clause can also be seen as part of a wider trend whereby the increased freedom of private property in the forests, established as noted under Catherine the Great, had begun to swing back during the later nineteenth century in the direction of increased state control. This trend was to culminate in Bolshevik forest legislation passed in 1918.

On the Eve of the First World War: Forest Protection by 1914

New forest codes issued in 1905 and 1913 sought to consolidate and, in the case of the 1913 code, to streamline the growing corpus of relevant law. Yet aside from adding somewhat to the responsibilities and powers of Forest Preservation Committees they did not introduce any significant new conservation developments. Thus, in 1914, D. M. Zaitsev, chair of the First All-Russian Congress of the Representatives of Forest Industry and Trade in St. Petersburg (convened specifically to examine a proposed, though never-enacted, further reform of existing legislation[56]), found the general shape of forest management and conservation—in law and in practice—essentially unchanged since 1888.[57]

---

[56] D. M. Zaitsev, *Lesnoe Zakonodatel'stvo (Kommentaryi k proektu novogo Ustava Lesnogo). Doklad 1-omu Vserossiiskomu S"ezdu Predstavitelei Lesnogo Promyshlennosti i Torgovli* (St. Petersburg: Biuro po uchrezhdeniiu S"ezdov Predstavitelei Lesnogo Promyshlennosti i Torgovli, 1914). Somewhat stricter forest conservation laws were in the process of being worked out in 1914, as part of an intended overhaul of the entire edifice of forest management. Zaitsev's comments were made in response to the proposed legislative changes. All plans were shelved, however, following the outbreak of war in 1914. Following the collapse of the tsarist regime in early 1917 forest reform re-emerged as a priority, though in somewhat more ambitious form.

[57] Though numerous amendments of the law had been decreed during the intervening period, these made few real changes. For example: "O Rasprostranenii polozhenii o sberezhenii lesov na gubernii tsarstva Pol'ska" and two subsequent similar decrees (issued 8 June, 1898, 9 July, 1899 and 23 March, 1901, respectively) extended the Statute's operation into Poland. These were areas in which *servituty* was practiced (see elsewhere in this chapter for an explanation of this system); "O merakh k okhraneniiu lesov Turganskoi oblasti ot istrebleniia," passed 26 February 1901, allowed for government purchase of certain types of forest land for protective ends (clause 1), and also increased the frequency and value of prizes given to owners or their workers who attained impressive results in forest planting. The "Statute" itself was reissued in 1901 to incorporate these and other alterations. Special instructions for the creation of

The area of Russian forest under the jurisdiction of the 1888 Statute had grown somewhat, however. Following Zaitsev's figures, by 1914, of the 55,516,243 *desiatinas* of forest now under the jurisdiction of the Forest Preservation Administration[58] (and thus of the Statute) 572,352—a little more than 1 %—had been designated protected. A further 733,364 *desiatinas*, about 1½ %, had been established as water-preservatory.[59] This brought the protected total to 2-2 ½ %. The rest—54,210,527 *desiatinas*—fell within the third category already noted. The entire 55,516,243 *desiatinas*, following Zaitsev's figures, were administered by sixty-seven Forest Preservation Committees comprising 657 individuals, only 147 of whom Zaitsev felt exercised any "real authority."[60] In practice, moreover, forest conservation enforcement fell to a bewildering variety of persons and groups beyond the Committees: local officials of the state, forest administration, foresters, and forest guards (of whom there were perhaps as many as 30,000 by 1914), local land captains [*zemskie nachalniki*] and even the general police.

How effective were these protections? Not very, according to Zaitsev, whose comments in this regard echo those of several other contemporaries. The problems were various; and though they were worst in the majority non-protected forests, they affected protected forests also. Thus, although state funds and other resources had been made available for establishing and maintaining both protected and water-preserving forests, those established on privately-held lands frequently suffered from uncooperative owners who, moreover, often could not

---

"protected forests" were issued in 1903, and subsequently. All amendments from 1888 to 1902 are collected together in N. I. Iakovlev, compiler, *Sbornik uzakonenii o sberezhenii lesov chastnykh i obshchestvennykh v guberniiakh evropeiskoi rossii i tsarstva pol'skogo* (St. Petersburg, 1902).

[58] This is perhaps between one-third and one-half of the total forest area of European Russian and a smaller fraction of the total within the whole Empire. The total amount of forest land under the Forest Department stood at 238.2 million in 1900 according to D. Giriaev, A. Savel'ev, "Kratkii ocherk istorii lesnogo khoziastva Rossii," *Lesnoe khoziastvo* 1 (1992). I have not been able to find figures for 1914, though they are likely to be quite similar.

[59] D. M. Zaitsev, *Lesnoe Zakonodatel'stvo*, 15.

[60] Ibid., 14. Of the 54,210,527 *desiatinas* under *sberezhenie*, 10,859,610 were at that time under exploitation in accordance with "forest economy plans" confirmed by local forest preservation committees, while the remaining majority—44,350,917 *desiatinas*—were, in Zaitsev's words, "exploited according to the inclinations of their owners, without any concrete adherence to forest preservation oversight."

be held accountable to existing legal protections, since these applied, "only in those cases where . . . cutting has been carried out after the confirmation of a [specific government] forest economy plan or in violation of a [specific] ban issued by the Forest Preservation Committee." "In and of itself," Zaitsev continued, "clear-cutting does not entail any [legal or criminal] responsibility on the part of the forest landowner, but simply gives the Forest Preservation Committee the right to prevent further such cutting in the relevant forest area for a period of time to be determined based on the degree of damage . . . or until a forestry plan has been worked out for that specific area."[61] Worse, based on some of the loopholes already noted, protections were routinely and legally over-ruled in cases involving the securing of borders, construction of roads, needs of agriculture, construction of villages, property division, "expansion of economically useful land at the expense of the forest economy," and so on.[62] Ultimately, Zaitsev claimed there was nothing really solid in existing law to prevent landowners from pursuing their own economic interests—at the expense of the state's—and converting forest land to pasture land, as was the main temptation.[63]

The problem was exacerbated by budgetary constraints, especially the very low pay scales for forest personnel, which acted as an invitation to bribery and a disincentive to enforcement.[64] The Forest Preservation Committees were,

---

[61] Ibid., 11, 13.

[62] Ibid., 45-46.

[63] Ibid., 49.

[64] Zaitsev noted in 1914 that the entire budget for forest conservation amounted only to 135,848 rubles, some 70% of which was spent on operating expenses. A mere 4,239 rubles was set aside specifically for the payment of individuals responsible for directly overseeing both private and public forests. Boreiko claims pay for guards (presumably during the 1880s) was 3-9 rubles a month versus 280 for their French counterparts. (V. E. Boreiko, "Plakala Sasha kak les vyrubili," 17). According to the nineteenth-century English forestry authority John Croumbie Brown, around 1880 the Commissioner of Woods and Forests [*Glavnoi Lesnik*] in a mining district in the Urals under whose jurisdiction lay thousands of square kilometers in Nizhni Novgorod, Kazan, Viatka, Perm, Tobolosk, Orenburg, and Ufa, received an annual salary of less than 800 rubles (100 pounds sterling). Under the Commissioner was no shortage of hired workers—under-foresters, woodsmen, and guards—but all, Brown noted, were proportionately under-paid, raising questions about their ability or desire to discharge their duties effectively. See John Croumbie

apparently, as susceptible as any in this regard. Thus, another forester, I. Gershman, writing three years earlier, in 1911, noted with some sadness that land speculators could typically "always get permission with little difficulty from the Forest Preservation Committee to clear-cut."[65] The truth of this statement can be seen in the many rulings and decrees issued during the First World War by Russia's numerous Forest Preservation Committees, especially those in European Russia's major agricultural regions. The great majority are little more than lists of cutting permissions, though some also refer, vaguely, to "forest economy plans" recently confirmed in this or that forest area.[66] Overall, Zaitsev concluded, "the activities of the forest preservation committees, in most cases, exist only on paper,"[67] while forest law in general remained a "typical bureaucratic product... powerless to protect the forests from destruction."[68] Consequently, he claimed, from 1887 to 1907, European Russia's forests had shrunk from 90,058,000 to 81,785,000 *desiatinas*,[69] a drop of some 8-9%.[70]

Bad as these figures sound, if the above-noted estimates are correct that

---

Brown, *Forestry in the Mining Districts of the Ural Mountains in Eastern Russia* (Edinburgh: Oliver and Boyd, etc., 1884), 124. According to Zaitsev, the Swiss, often cited at the time as paragons of forest management, spent more than one hundred times as much per *desiatina* on their forests as Russia—35 kopecks against about 0.24 kopecks in Russia (D. M. Zaitsev, *Lesnoe Zakonodatel'stvo*, 29).

[65] I. Gershman, "Ocherk istorii lesovladeniia," 516.

[66] The journal *Lesnaia zhizn i khoziaistvo*, among others, published many dozens of such (concerning the Tambov region in this journal's case) during 1916-1917. For a representative example, see 1 (1917): 50-52.

[67] D. M. Zaitsev, *Lesnoe Zakonodatel'stvo*, 14.

[68] Under such conditions, violations of forest law were frequent and serious while enforcement was rare and gentle. For example, in *Yaroslavskaia Guberniia* in 1908 only one case against violators of forest law was brought and this sought punishment only in the amount of 13 rubles. The success of the prosecution is not indicated. Nor is it at all clear whether this example is typical or exceptional (ibid., 13-15.)

[69] Ibid., 16.

[70] V. E. Boreiko gives an almost identical figure for the period 1888-1908, during which years, he claims, in European Russia minus Archangel and Vologda provinces some 9 million hectares were destroyed. See, V. E. Boreiko, "Plakala Sasha kak les vyrubili.," 16.

between 1867 and 1914 as much as 30% of the entire forest area of European Russia was cut down, and in the absence of figures or any other evidence attesting to a marked increase from 1907-1914, this would actually suggest that the greatest destruction occurred during the two decades *prior* to the issuance of the "Statute on the Safeguarding of Forests." Such a statement, however, must remain speculative, given the paucity of figures in general, a lack of clear indication that they encompass identical territorial expanses, and other factors. (There is little question, however, that forest destruction would pick up handily *after* 1914, at least for a few years.)

The overall reduction of forest area during the nineteenth century, however, hid considerable regional variations. In Iaroslav province, for example, half of all forests disappeared during the century from 1774-1874. In Kharkov and Ekaterinoslav provinces a full 80% fell to axe or flame in a single decade from 1872-1882. In Kiev province from 1849-1869 187,000 *desiatina*s were cleared.[71] Further north and east the damage was considerably less.

The problem of corruption is illustrated quite dramatically in the writings of the English forester John Croumbie Brown, who travelled in Russia just before the passage of the 1888 Statute. His comments would remain valid for decades to come, moreover. They are worth relating at some length here since they also illuminate local culture and practices with which any attempt at organizing forest use would have had to contend, and which were to be problematical also for the Bolsheviks, whose organizational ambitions were especially great. By law, Brown reported, the average *zavod* (mill or plant) was at that time entitled to use a maximum of "300 cubic fathoms of wood a year from Government forests for the production of steam." Such timber could not legally be cut randomly. Procurement areas were designated by the local authorities—in accordance at that time with the terms of the 1882 decree noted earlier. Not infrequently, these might be situated a good distance from the *zavod*, significantly increasing the cost to the user. In practice, however neither the limit on timber cut nor the local authorities' directions to certain forest areas were always complied with. Croumbie states:

> I was informed on credible authority that if . . . there were handed to the inspector, along with the permit, a sum of money enclosed in an envelope, much more satisfactory arrangements might be made. And if the amount enclosed amounted to 200 rubles . . . arrangements might be made for the wood . . . [to be] cut in forests within a tenth of the [required] . . . distance; but then all the officials who were under the

[71] Ibid. These figures, also, suggest that under the tsars, at least, destruction peaked prior to 1888.

> [bribed] inspector must also be silenced by hush money, as otherwise some fatal objection might be raised. The forest guard might be satisfied with 20 or 30 kopecs . . . but his superiors must receive much more, according to their rank; and then though 1,000 cubic feet might be felled and delivered, in the official records everything would be found in order—300 cubic feet granted, 300 cubic feet felled, 300 cubic feet delivered.[72]

Similar arrangements also operated in the case of domestic use by village communes, each of which would also typically have a certain amount of forest allotted to it.

> [T]he quantity of firewood to be allowed for heating and cooking seems to have been left to the discretion of the Chief Commissioner of the Imperial Forests; and you have to address and present to him a petition for authoritative permission to cut down the quantity allowed to you —he having a list of those persons who have had the privilege granted to them. Moreover, you have to state in your petition how many houses, rooms, and fire-grates you have requiring fuel . . . . When your petition is presented, if it be unaccompanied by something to make it presentable [a bribe], it will be a long time before it is answered, and you may long dance attendance before there is time to attend to you. There is such an awful press of business—so many to be attended to before your turn comes! The proper authorities are not at home, gone to other districts, cannot tell when will be best; and when it does come your quittance is numbered for the most distant estate in the boundary of your town, perhaps 40 or 60 versts [25 or 37 ½ miles] off, instead of 8 or 12, and what are you to do? What can you do?"[73]

Even "honest" officials could apparently be got around in similar fashion. In one of the more comical of his numerous anecdotes Brown relates the substance of an occurrence—hypothetical, though supposedly typical—in which just such a man is bought off anyway via his wife, who has herself been bribed and successfully persuades her husband to acquiesce in the case against her briber.[74] Persons wishing to avoid all these expenses and bothers could, of course, always make arrangements with any of the "numerous forest robbers" or "wood stealers" in operation around the country. Alternatively, if one's needs were modest it was usually worth the effort simply to cut down a tree or two oneself. "I think," Croumbie's source concluded,

---

[72] John Croumbie Brown, *Forestry in the Mining Districts of the Ural Mountains in Eastern Russia*, 126.

[73] Ibid., 126-27.

[74] Ibid., 129.

> You will see . . . that it is no wonder the forests are rapidly disappearing in all parts of Russia, especially when you consider the careless humble-jumble 'self first' system . . ., and besides this, the want of proper scientific principle honestly carried out. I believe matters are somewhat better than they were, but it is too late locking the stable door when the horse is gone. And I am of the opinion from what I have seen wherever I have been, that since the introduction of steam power the consumption of firewood is very much more rapid than the growth, for they are now cutting down young timbers, where there are any to cut, in a shameful manner which is heart-rending to see.[75]

Like Zaitsev some thirty years after, Brown considered these various corruptions and breakdowns symptomatic of the relatively low pay given to all levels of forest personnel.[76] Unlike Zaitsev, however, he also blamed them in part on Russian culture.[77]

Comments such as these focus largely on the negative role played by local entrepreneurs and private landowners intent on maximizing profit, and on rural populations largely fulfilling their own local and personal needs. To be sure, these groups played a pivotal role in the decline of Russia's forests in the last century of Imperial rule and attracted the preponderance of attention from

---

[75] Ibid., 133-34.

[76] D. M. Zaitsev, *Lesnoe Zakonodatel'stvo*, 14.

[77] Croumbie Brown's rather generalistic descriptions of corruption during the late nineteenth century do not, of course, constitute a revelation, and are supported by various other sources for that period and later. One worthy of very brief note, perhaps, for its close agreement with the practices earlier described, involves the testimony given in August 1917 by Mikhail Andreevich Ivanov, a forester [*lesnichii*] against a co-worker. Empowered by the Temporary Forest Executive Committee [*Vremennyi Lesnoi Ispolnitel'nyi Komitet*] under the Provisional Government to act "as an intermediary [*posrednik*] in connection with the transfer of the Severnoborskii forest area [*lesnichestvo*]" from the jurisdiction of a forester by the name of Beinar to that of another, and also to "ascertain the attitude of the population" toward the activities of the same Beinar. Ivanov's report is a litany of well-documented complaints and accusations of repeated corruption against Beiner, for issuing numerous "fake cutting permits," pocketing fees paid for them, falsifying records, and so on. Ivanov also accused Beinar of negligence in connection with a fire in the Severnoborskii forest region which burned from 7 April to 9 May, 1917 and as "a result of which a huge parcel of forest was destroyed completely and needlessly, since with assistance from the forester [Beinar] the forest could have been saved." *Gosudarstvennyi arkhiv rossisskoi federatsii* (hereafter, GARF). Fond 5520, op. 1, del. 34, st.st. 74-76.

advocates of major forest reform, as will be seen in more detail below. On the other hand, the state too was a major and growing consumer of forest materials. The Ministry of Transport (*Ministerstvo putei soobshcheniia*) and the Naval Department, (*Morskoe vedomstvo*) in particular required huge amounts of timber for railroad and ship construction, and for fuel. And timber export constituted a significant part of the Russian economy overall, as already noted. On the other hand, there seems to have been a widespread consensus that exploitation of state forests was much better organized than private forestry, and that it was carried out in a somewhat more sustainable manner. In fact, of course, the lines between state and private forest use were neither clear nor absolute. Private entrepreneurs acted typically under license from the state; and the state, in the form of the organs already noted and others besides, was a primary customer of private lumberers.[78]

During the twenty years or so immediately prior to the Bolshevik Revolution, control over state exploitation of forests came increasingly into the hands of the Forest Department itself (a trend which was later to reverse, see chapters 5-7). This was not an activity for which the Department—essentially an administrative, not a procurement organ—had originally been created. Nonetheless, it willingly undertook these new obligations beginning around 1898 at the request of other government organs, chiefly the Naval Department and the Ministry of Transport. Rising demand for timber, the escalating cost thereof, and bureaucratic and other difficulties involved in procurement inspired these organs repeatedly to approach the Forest Department with orders for materials (from state land), which it agreed to take on. In addition to cutting and supplying timber for these and other organs, over the following decade the Forest Department expanded also to serve peasant communal customers, further underscoring the many interconnections among private, state, and communal forest users. In the course of these activities the Department also set about acquiring, establishing and running a growing number of sawmills, and began to operate its own flotilla of ships and barges to move forest materials about the country.[79] Nonetheless, relatively little attention was given to these activities by would-be forest conservationists, who focused on privateers and poachers.

---

[78] See, for example, N. I. Faleev, "Kapitalisticheskye osnovy lesnogo khoziaistva," 1-2.

[79] GARF. Fond 5520, op. 1, del. 34. st.st. 13-20. (This is an untitled and anonymous article, the last pages of which are missing.)

Tsarist Forest Protection: In Retrospect

Given the various problems and negative assessments of Imperial forest protection already noted, it would be easy to concur with the dismissive attitude of Soviet authors and others who note the pre-Revolutionary period only in passing and by way of contrast with the supposedly more substantial achievements of later times. On the other hand, there are several reasons for a somewhat more optimistic reading. (The rather notable failures of early Soviet forestry, which will form the subject matter of later chapters, are at least one such reason.) First, as previously suggested, it is apparent—though not absolutely clear—that the worst periods of forest clearing occurred during the decades immediately prior to the passage of the "Statute on the Safeguarding of Forests" in 1888. Second, whatever their shortcomings, the protections put in place that year were a clear improvement over those obtaining before, even if they did not prove sufficient to stem the overall onslaught on Russia's forests. That clear-cutting could be stopped at all—even if after, rather than before it had been begun—was of some value, for example, as were the more effective legal protections made possible in the small quantity of protected forests. If the war was going badly, so to speak, existing legislation, when enforced, proved sufficient at least to win a few battles. Thus, some contemporary assessments of the 1888 law were, in fact, quite positive. For example, writing in July 1917 from first-hand experience gathered over many years, Aleksei S., a member of The Foresters' Union (*Soiuz lesovodov*, or *Soles*, founded in April 1917 to unite forest workers of all rank and type) credited the Statute with bringing improved order and protection to the formerly "more or less chaotic state" of forestry in the Iuzhnoberezhskoe *lesnichestvo*[80] in the Crimea. "Up to the early 1890s," he wrote, "public forests" including "those assigned [to the use of factories], and . . . state forests were being destroyed by fires, over-cutting and the grazing of cattle." A huge blaze during 1881-1883 in particular had wiped out the "whole lower band of deciduous forest." Despite local bans issued back in the 1840s, the local population's insistence on using the fire-cleared and other forest areas for grazing had made regeneration impossible. The 1888 Statute, despite never having been well designed for the unique forest conditions of the Crimea itself, nonetheless brought "a significant portion of good." Certain forests in the area "were [as a

[80] *Lesnichestvo*—"forest area"—is the usual official or professional designation for a unit of forest, especially within a forest plan.

result of the law] declared 'protected' in 1890," following which the forest administration "was given the right to insist on vital . . . limitations in these forests on both cutting and grazing." Over the complaints of some elements of the population (a sign protections were in fact somewhat successful) at least two local *zemstvos* enforced the new rules with "an iron hand," achieving, if this report is to be believed, "strict protection in state forests and in those assigned [to industrial plants] from cutting and seizure," as well as better general distribution of forest products to local peasant communes, and increased revenues for the state.[81]

Another positive assessment of the same law was given in an article published in *Lesnoi zhurnal* (The Forest Journal) in 1911, which asserts that despite other deficiencies it brought about a perceptible decline (in European Russia overall) in the practice whereby forest lands purchased by speculators from impoverished landlords were being clear-cut for quick profit by the new owners.[82] In this direction it might also be noted that based in part on the "Statute" real achievements were made prior to the Revolution in artificial reforestation also, especially on state forests where between 1842 and 1914 899,000 hectares of new forest were planted.[83]

Other Aspects of Nineteenth-Century Forest Protection: Fires, Insects

Alongside general forest protection legislation of the type already noted, indeed intersecting with it, tsarist Russia also passed decrees designed to combat other, more specific forest threats, especially fires and insect infestations, both of which could have natural, anthropogenic, or combined causes; and both of which were surprisingly widespread and frequent. Here too, both the efforts and the results are clearly more than negligible, even if the odds were always heavily against those seeking to protect.[84]

---

[81] Aleksei S. (author's full last name is illegible). "Otnoshenie iuzhnoberezhskogo lesnichego k naseleniiu i lesam za poslednie 20 let/1898-1917." Deposited in GARF 5520, op. 2. del. 18, st.st. 112-14.

[82] I. Gershman, "Ocherk istorii lesovladeniia," 516. *Lesnoi zhurnal* was founded in 1833.

[83] *Lesnoe Khoziaistvo SSSR za 50 let, 1917-1967*, 15.

[84] An article in the forestry journal *Lesopromyshlennyi vestnik*, 42 (1912), 540, claims that from 1895 to 1909 in the sixty-three provinces of European Russia there were no fewer than one million forest fires burning some 2.8 million trees down.

The fire-fighting provisions of nineteenth-century Russia's forest conservation legislation date at least to a law passed 9 August, 1800 by the Emperor Paul requiring all serfs within 25 *verst*s of a fire (on all categories of forest land) to assist in extinguishing it and to this end to bring with them axes, spades and other personal items. Strict penalties were established for non-compliance.[85] This regime was impacted on some lands by the promulgation a little more than a year later—on 11 December, 1801—of a decree by Paul's successor Alexander I granting the right to all free Russians to own unsettled (including forest) land. On such land, responsibility for preventing and fighting fires—and for mustering peasant aid in this cause—now fell on the landowner himself, a responsibility which was not always met or easily enforced. Recruitment of laborers for such work was often harder still to ensure. State efforts to draft forest conservation workers have already been noted, as have the problems presented in this regard by the emancipation of Russias millions of serfs in 1861. After this date, and further in search of solutions to forest labor shortages the state increasingly adopted a system of monetary rewards offered to peasants in return for their services.[86] At first unregulated and entirely ad hoc, this payment system was gradually made mandatory. In 1867 the state began to require payment for all peasants coming from a distance of greater than fifteen

[85] Gosudarstvennyi sovet, *Stenograficheskie otchety* (St. Petersburg: Gosudarstvennaia tipografiia, 1908), Sessiia IV. Zasedanie 5 (3 May, 1908), 92. Evidence exists that fires were set deliberately for a variety of purposes: by peasants set them seeking pay or other rewards for subsequently helping put them out; by individuals hoping to get around forest protection laws or fearing the loss of timber to state expropriation and especially in connection with the land reforms carried out in the immediate pre-Revolutionary period by Prime Minister Peter Stolypin; to clear land for grazing; to drive animals when hunting; and so on. More information is given in some of these regards elsewhere in the text. Insect infestations were also sometimes deliberate. V. E. Boreiko, for example, relates that in the early twentieth century at least it was not uncommon for harmful beetles or larvae to be deliberately placed in "two or three" trees, from which base they would rapidly spread. Those responsible would then notify the local Forest Preservation Committee, which would be forced to condemn the area and order it clear-cut and burned to protect neighboring areas. The fear of such infestations was sufficiently great that in some cases a simple announcement that one's land was infected, even if it in fact was not, was enough to gain permission to clear cut. See, V. E. Boreiko, "Plakala Sasha kak les vyrubili," 18.

[86] Gosudarstvennyi sovet, *Stenograficheskie otchety* (St. Petersburg: Gosudarstvennaia tipografiia, 1908). Sessiia IV. Zasedanie 5 (3 May, 1908), 91-94.

versts to help fight fires in state forests. This law was extended in 1872 to cover private forests also. In 1908, following a lengthy debate in the State Council, payments were made mandatory for all categories of forest and regardless of the distance a peasant had travelled. Up to the beginning of the First World War the Russian government continued to allocate funds for these payments.[87] The concerns of some lawmakers—that rewards would only encourage the deliberate setting of fires by peasants seeking payment for extinguishing them—seem to have been outweighed by concerns, backed up by repeated observation, that the main risk was that peasants typically tended not to show up at all, or to work for the welfare of their forests in any way, in the absence of some material reward.[88] These developments aside, Paul's law remained technically in force throughout the period up to 1908. Extinguishing fires on peasant communes was the responsibility of the commune itself.

Protecting forests from damage inflicted by insects and their larvae also received at least some attention well prior to 1917, though more in the realm of research than of action. A two-volume manual, *On Harmful Insects* (*O vrednykh nasekomykh*) was published in 1845-1851, helping lay the foundations for a scientific approach to the problem. From 1887-1898 one of the founders of Russian forest entomology, I. Ia. Shevyrev published a major work that served as a basic text for the next several decades. At the turn of the century an entomological laboratory was established within the Forest Department and charged with studying the impact of insects on forests, methods of protection, and so on. At about the same time a bureau of mycology and phytopathology (the study of funguses and plants, respectively) was set up within the Ministry of Agriculture.[89] A growing, if small, number of specialists continued to study these problems and to carry out relevant experimental work up to the Revolution and beyond. The draft of a revised forest conservation law debated on the eve of WWI by Zaitsev and others at the First All-Russian Congress of the Representatives of Forest Industry and Trade in St. Petersburg (discussed above)

[87] Legislation voting funds to cover expenses thus incurred during 1912 was passed the following year. See ibid. (1913). Sessiia VIII. Zasedanie 33 (8 May), 1664.

[88] For this debate, see ibid. (1908). Sessiia III. Zasedanie 35 (11 May), 1630-38; Zasedanie 41 (27 May), 2175-79; and Sessiia IV. Zasedanie 5 (3 October), 91-112. The law of 1908 referred to above was signed by the tsar on 4 December.

[89] *Lesnoe Khoziaistvo SSSR za 50 let, 1917-1967*, 152.

included provisions empowering provincial land assemblies to "compile compulsory regulations" designed to extend "full protection" to "useful"—i.e. insectivorous—birds "with the goal of eradicating harmful insects in the forests."[90] The benevolent role of certain birds in this regard was debated frequently both before and after the Revolution, though legislation designed to offer protection even to those birds deemed of particular importance did not emerge at any point under consideration in this study.[91]

Though major initiatives to eradicate targeted insect pests, typically in the dubious form of pesticide bombardments, came only after 1917, several "successful experiments" in protecting forests from pests were carried out before the Revolution by V. N. Stark. P. A. Polozhentsev researched into the capacity of trees to withstand certain types of infestation. Soviet sources note also that "measures" for protecting forest areas artificially planted on the steppe against insects were "worked out" prior to the Revolution, though it is unclear what these measures were or if they were effective.[92] Nonetheless, taken together, these various approaches to many aspects of forest protection suggest a relatively active, if not a spectacularly successful, pre-Revolutionary conservation tradition.

Toward Building a Consensus for Forest Reform: Professional Foresters Speak, 1905-1917

During the decade prior to the First World War, the various inadequacies of Russian forestry—from inefficient modes of exploitation to highly unsatisfactory levels of preservation—inspired voluminous discussion on reform within major forest congresses, societies, in forestry journals, and in other media. Within these debates, and certainly regarding forest protection, a few specific themes began to be stressed over and again, to the point that it becomes possible to discern the parameters of a rough consensus—shared by an apparent majority of foresters—on what problems existed, and, to only a slightly lesser degree, on

---

[90] D. M. Zaitsev, *Lesnoe* Zakonodatel'stvo, 21-22.

[91] There were, of course, indirect protections provided to birds through the establishment of nature preserves (*zapovedniki*), which provided safe havens, on paper at least, to all creatures within their borders. And hunting laws offered some level of protection to a few specific bird species, though not the insectivores under consideration here.

[92] *Lesnoe Khoziaistvo SSSR za 50 let, 1917-1967*, 267.

their solutions. In some cases this consensus is easily quantifiable, such as in the case of conservation propositions raised and voted on at specific forest congresses, or the frequency of published articles making similar points. In other cases, the existence of general trends among foresters can only be inferred from evidence such as the tone and content of dialogue, and so on.

At the same time, beyond the circle of academic and institutional foresters, popular attitudes toward the forests, and toward matters of their exploitation and protection, also show certain general, consensual, features. These latter were generally of considerable vintage, in large part constituting or drawing upon the accumulated folk-wisdom of many generations of peasants living in or by—and directly depending upon—Russia's forests. As will be seen, though much overlap existed between the basic ideas of professionals and peasants there were also fundamental differences, particularly regarding questions of central or local control, the need for overall planning and scientific management, and the role of experts. Following the collapse of the tsarist regime in 1917—which essentially opened the field for would-be forest reformers of all types—these questions, previously somewhat academic and muted, were to become fiercely-contested battlelines (see chapter 3). A discussion of popular notions relevant to forests, however, will be deferred to the next chapter, where they assume central importance. Here, it will be appropriate to consider the major themes for reform proposed by professional foresters, and which were to provide much of the basic framework for legislative activity up to and, significantly, well beyond 1917.

As already noted, a constant theme to debate on forestry reform, and especially regarding forest protection, was the widespread and increasingly bitter antipathy toward private forest ownership. "Is it any secret," asked a member of the Forest Section of the First Kiev Agricultural Congress in 1913, striking a popular note at that time, that the majority of our private forest-owners are liquidating their forests, without any use at all of educated economic calculations, and that following this in many places after a few years there begin to emerge exhausted areas, no longer repaying fully the heavy peasant labor put into them, and that in poor years there is no cheap fuel, no winter supplies, no materials for repairs or restoration of burned buildings to be had at accessible prices?[93] Behind comments such as these lay a widely-held perception that over the past century or so a demographic shift in forest ownership had transpired, characterized by

[93] "Lesnaia sektsii na pervom kievskom sel'skokhoziaistvennom s"ezde i printsipy obshchego lesookhraneniia," *Lesnoi zhurnal* 9-10 (1913), 1434.

reduced noble and increased merchant capitalist holdings, and which had led to wider and more rapacious cutting.

The rise of non-noble private proprietorship—and a corresponding decline of noble ownership—was a clear and widespread trend (see table 3). Whether or not the new class of owners was indeed far more exploitative of their holdings than its predecessor, and thus a much greater threat to the forests in general than the traditional landlords, is harder to assess, however, though the general rise in cutting over the second half of the nineteenth century is itself clearly established. One observer, I. Gershman (whose modestly positive assessment of the 1888 "Statute" has already been noted) asserted in 1911 that a chief motivation of the private purchaser and speculator in forest land was to clear-cut in order to profit both from the sale of timber and ultimately of the land itself, or from the sale of agricultural or other goods to be produced on the cleared land.[94] The process was, he estimated, inextricably linked with the wider process of industrial expansion, which though in Russia far from an affair of exclusively private capital,[95] certainly provided opportunities for the private speculator: "Every new railroad, plans for which forest industrialists [*lesopromyshlenniki*] find out about, provokes speculative purchasing of forests in the area where the railroad is going to be laid."[96] Another, V. Rostovstev, noted in 1913 that the "insufficiencies of forest law" were being exploited on privately-owned forests much more than on their state counterparts. "Almost exclusively on the latter alone," he asserted, was it still possible find trees of "good size."[97]

---

[94] I. Gershman, "Ocherk istorii lesovladeniia," 515-16.

[95] European industrialization was financed primarily by private capital in Britain, by a mixture of state and private funds in Germany, and mostly by state investment in Russia, creating a general picture of diminishing private and growing state influence the further east one looks. See, for example, Alexander Gershenkron, *Economic Backwardness in Historical Perspective: A Book of Essays* (Cambridge, MA: Belknap Press of Harvard University Press, 1962).

[96] I. Gershman, "Ocherk istorii lesovladeniia," 516. Interestingly, Gershman noted that the growth of this kind of "speculative" (*kupecheskyi*) forest ownership—and thus of forest misuse—was least in those western provinces "where Jews are forbidden to purchase land" (517) implying thereby, a link between Jews and forest destruction.

[97] V. Rostovstev, "Lesokul'turnye raboty v Kalenkovicheskoi dache, Kalenkovicheskogo lesnichestva, Minskoi guburnii," *Lesnoi zhurnal* 8 (October 1913): 1206.

Table 3: Patterns of Private vs. Noble Landholding by Province, 1877 and 1905 [98]

Province	% Merchant Land 1877	% Merchant Land in 1905	% Noble Land 1877	% Noble Land in 1905
Arkhangel	0.1	0.5	0.1	0.0
Perm	12.9	11.7	48.3	39.1
Novgorod	12.3	19.1	37.6	18.5
Iaroslavl	2.9	6.2	29.3	14.5
Moscow	8.4	11.9	27.8	17.4
Smolensk	6.1	13.3	44.0	21.0
Kurlandia	1.3	2.3	49.6	48.2
Kovno	0.3	0.2	48.6	43.5
Vologda	3.5	8.1	17.3	4.8
Viatka	2.6	4.5	7.0	3.3
Pskov	5.2	12.6	37.4	17.7
Kostroma	10.7	16.5	38.9	18.7
Nizhnyi Novgorod	2.7	7.6	35.1	19.6
Estland	2.3	3.4	53.5	68.2
Vitebsk	3.4	7.1	49.9	34.7

[98] Adapted from I. Gershman, "Ocherk istorii lesovladenii," 517. The rise in non-noble proprietorship was rather uniformly widespread. This said, some of the strongest non-noble gains came in heavily-forested northern regions, such as Arkhangel (a 400% increase), and around Moscow (including Vologda and Nizhnyi Novgorod at well over 100% each). This lends at least some credence to the noted sense among forest specialists of a considerable rise in private forest ownership.

Table 3 (Continued)

Province	% Merchant Land 1877	% Merchant Land in 1905	% Noble Land 1877	% Noble Land in 1905
Volyn'	2.8	3.5	49.0	39.9
Olonets	--	8.2	9.5	1.3
Ufa	4.8	4.8	24.4	14.6
St. Petersburg	9.4	14.7	48.2	33.4
Tver	4.2	6.1	27.7	13.0
Vladimir	8.1	10.2	29.9	17.0
Livland	1.0	0.1	48.6	45.7
Mogilev	2.1	2.3	53.2	35.7

To meet the challenges of destructive private forestry—and as a necessary prerequisite for better forest protection overall—foresters again and again emphasized the need for greater state control. Indeed, this is without doubt the single most consensual general aspect to pre-Revolutionary forest protection debate.[99] A variety of opinion existed, however, regarding what this meant more

[99] There were, of course, numbers of individuals who vehemently opposed any idea of an increase in government activity in the forests or any tampering with private property rights. Overall however, these individuals typically can not be considered conservationists or reformers at all, since they rarely if ever even spoke explicitly on this matter. In writing throughout this chapter of a kind of "consensus" among foresters, I have in mind only that majority of foresters who called for better forest protection. Among these I have found no evidence of clear opposition to bigger government, though as detailed in the text, the nature and degree of governmental activity in the forests they envisioned varied considerably. An exception of some sort to pro-conservation foresters' desire for a greater government role in forestry was the occasionally-mooted idea that there should be a tax reduction or abatement, land taxes had actually been higher on forest land, encouraging cutting and drawing criticism from many foresters.

specifically. One relatively popular idea involved increasing the jurisdiction of existing legislation—which stemming from 1888's "Statute on the Safeguarding of Forests," applied only to a large part of European Russia—to cover every forest in the entire Russian Empire.[100] Forest congresses in 1911 and 1913 in St. Petersburg and Kiev, respectively, debated this very question.[101] Arguments against expanding the law focused in both cases on a few key issues, many of which boil down essentially to an argument about whether forest was Russia's wealth or a contributor to its relative poverty. In support of the latter, it was argued that Russia had too much forest, which drove down the price of timber and wood fuel and made a modernizing transition to brick and cement, oil and coal, economically infeasible; that agricultural land had more economic value than forest land and was in relatively short supply against an overabundance of forest; that the climate-protecting qualities of forest were poorly understood, often exaggerated, and thus not a valid basis for conservation[102]; that western Europe had few forests but better-developed economies.[103] Conservation, these men argued, might be improved where necessary, perhaps, but it should not be initiated elsewhere. These sentiments, however, lost out in final voting in both 1911 and 1913 to the advocates of expanded conservation,[104] who in 1911 approved the following resolution: "The Forest Section has come to the conclusion that protection must be widened to [cover] all Russia's forests" with

[100] Only forests were envisioned here. There was no sense of any need to include also what were sometimes called "insignificant stands," "very small woods," and so on.

[101] The *Vserossiiskii vneocherednyi s"ezd lesovladel'stev i lesokhoziaev dlia obsuzhdeniia lesookhranitel'nogo zakona* took place in St. Petersburg from 23-27 January, 1911. In 1913 the Forest Section of the First Kiev Agricultural Congress took up the same question.

[102] This is, of course, exactly the objection raised in the present day by opponents of legislation designed to curb global warming by capping greenhouse gas emissions.

[103] See for example, V. F. Kliuchnikov, "Konkretnye zadachi lesookhraneniia," 1410-24.

[104] At the 1911 St. Petersburg Conference, expansion of forest conservation to cover all the Russian Empire was carried by a slim margin of 133 votes to 128. "Lesnaia sektsiia na pervom Kievskom Sel'skokhoziaistvennom S"ezde i printsip obshchego lesookhraneniia," *Lesnoi zhurnal*, 1-2 (1910), 1368. The Kiev congress was a relatively large one, with 933 persons in attendance. Its Forest Section, however, was much smaller, consisting of 30 persons. (Ibid., 1367.)

the exception "only of the smallest forest parcels which cannot constitute an object of proper forest management and which do not have any protective, water-preservatory or any other particular significance."[105] The same Congress also moved to "accept in principle and work out formally principles of actual state-public guardianship over [all] private forests."[106] These proposals were re-read and supported again at the 1913 Kiev conference.[107] In fact, though much ink and spit was to be spent drafting and debating major reform (especially during 1914, see above on D. M. Zaitsev and the First All-Russian Congress of the Representatives of Forest Industry and Trade in St. Petersburg), no significant changes were to be made to forest legislation or its jurisdiction from this point until after the collapse of the tsarist government in early 1917.

Beyond calls for expanded government *control* of forests, there was also strong support for more state *ownership*. Often this went no further than advocacy of continued or increased expropriation of uncooperative or abusive private or communal owners, building on the practice already established in law since 1888. This was the case, for example, when several participants at a 1905 meeting of the St. Petersburg Forest Society spoke out in favor of "obtaining for the state, [more] peasant and privately-owned forests."[108] Suggestions were also made here and elsewhere for the creation of a fund to finance these purchases, though with

---

[105] Ibid., 1377. The Resolution was voted for 10 to 7.

[106] "Lesnaia sektsiia na pervom kievskom sel'skokhoziaistvennom s"ezde i printsip obshchego lesookhraneniia" (continuation), *Lesnoi zhurnal*, 9-10 (1913): 1436.

[107] Ibid., 1426-1428 and *passim*. Although the Congress voted support for the "necessity of more widespread forest protection or, better, [for] care [*opeka*] of all the forests of Russia" it did not do so unanimously but by a relatively slim 10-7 majority, with 13 abstentions. Voting on the same question two years earlier at the St. Petersburg congress produced similar results, with 31% supporting increased protection against 29% opposing and 40% abstaining. In both cases it is unclear what, if any, specific measures were under discussion, nor why votes were cast as they were. In any case, these figures argue against simplistic acceptance of the overwhelming popularity of increased forest protection among all involved parties. See, ibid., 1429.

[108] "Protokol' 9-go ocherednogo zasedaniia S. Peterburgskogo lesnogo obshchestva, 26 November 1905," *Lesnoi zhurnal* 2 (1906): 188.

little apparent success.[109] A decree of the Extraordinary All-Russian Congress of Forest Owners and Forest Managers (*vneocherednyi vserossiiski s'ezd lesovladel'tsev i lesokhoziaev*) convened especially to discuss forest conservation law, though understandably lacking expropriationist fervor, did suggest the "necessity" of "obtaining as state property forests owned by peasant communes if in these there is not being carried on proper [*pravil'nyi*] forest management." Even regarding their own forests, however, participants sometimes espoused surprisingly pro-regulatory views. For example, not only was there general support for increased direct government *assistance* — primarily of a technical nature, in helping them improve their forest practices, especially forest regeneration[110]—even more noteworthy, the congress went so far as to accept the need for expropriation of *privately-* as well as communally-held land in cases where the owner refused to work in accordance with a legitimately established "economic plan" for his holdings (such plans to be worked out in agreement with the landholder in the first place, of course, as established in 1888). In contrast to the various comments made by numerous foresters before and after 1917 that private forestry was the least sustainable, and the least amenable to benign forest practices, the congress under consideration here provides considerable evidence of the interest of many private forest owners in better forest management, including conservation, which they perhaps saw as to their own private benefit as well as to society's more broadly. In this spirit, they resolved also for establishing more clearly in law the conditions under which clearcutting was not to be permitted, for a better definition of "clear-cutting itself," for the preservation of "useful" (i.e., insectivorous) birds, and so on.[111] The Congress also resolved in favor of putting into practice plans to create forest and other "zapovedniki" in accordance with the forward-looking—and ultimately highly important—ideas of

---

[109] M. M. Orlov, for example, suggested in 1911 a 5% contribution toward such a fund from profits made in state forests. The need to use such a fund specifically for government purchases of private forests was raised again in the Kiev congress during 1913. See, "Lesnaia sektsiia na pervom kievskom sel'skokhoziaistvennom s"ezde i printsip obshchego lesookhraneniia," *Lesnoi zhurnal* 9-10 (1913): 1435.

[110] "Postanovlenye, vynesennye vneocherednym Vserossisskym S"ezdom lesovladel'tsev i lesokhoziaev dlia obsuzdeniia lesookhranitel'nogo zakona." Held in St. Petersburg 23-27 January, 1911. *Lesnoi zhurnal* 3-4 (1911), 377-78.

[111] Ibid., 377-81, 387.

I. P. Borodin,[112] whose central role both in Russian conservation and its historiography are outlined further below.

Numerous other examples exist of calls for expanded state ownership. Zaitsev, for example, in the course of his 1914 assessment of existing and proposed forest legislation, pushed the envelope a little further, noting that forest protection in general was "the business not of private individuals, but of the government." The state, he continued, should buy up from their current owners "at least all protected and water-preservatory forests"—as well as "marginal lands" suitable for artificial replanting.[113] This was far from the only such suggestion.

Still more radical calls, not for limited transfers but for outright state ownership of *all* forests were also mooted at least as early as 1905. That year, for example, at a meeting of the St. Petersburg Forest Society, M. M. Orlov called for "concentrating [all] the forests in the ownership of the state" and G. N. Vysotskii—one of Russia's most illustrious silviculturists[114]—made a similarly sweeping proposal, differing from Orlov only in focusing on *zemstvo*s (local organs of government dating from a major reform of 1864) rather than the central state apparatus as the appropriate vessel in which to place control of this proposed national inheritance.[115] Similarly, in 1907 the journal *Lesnoi zhurnal* published an article on Russian forest protection by the German Social Democrat Georg Fol'mar (Vollmar) stating that all Russia's forests "must be in state ownership" and organized as a single entity.[116] Any efforts that might be made to combine both private and state ownership or management he castigated as "an agreement

---

[112] Ibid., 381.

[113] D. M. Zaitsev, *Lesnoe Zakonodatel'stvo*, 57.

[114] Vysotskii (1865-1940) is a well-known figures in the history of botany, forestry, and plant geography in general. He did pioneering research on the relationship between forests and climate, and was the first to show why there were no forests on the steppe.

[115] "Protokol' 9-go ocherednogo zasedaniia S. Peterburgskogo lesnogo obshchestva, 26 November 1905," 191, 203, and *passim*.

[116] Georg Fol'mar, "Sovremmenoe sostoianie voprosa o lesookhranenii," *Lesnoi zhurnal* 7 (1907), 1100-01. Fol'mar/Vollmar (1850-1922) was an important figure in the reformist socialist movement in Germany and more widely. His interests in rural Russian matters extended beyond forests to include peasant politics, where he championed the kulaks as an important social class vital to progressive reform and improved agriculture.

of two people who have caught . . . a hare . . . . One wants to kill it, the other to let it live. After quick reflection they cut it in two so that one might eat his half and the other let his live."[117] Opponents of state-owned forestry, Fol'mar continued, themselves supported and benefited from government ownership and control in industry and railroad expansion, rendering these persons' forest-related complaints somewhat hypocritical. The goals of even the least socialist-minded foresters for more efficient and uniform modes of forest exploitation and protection, he added, led logically toward socialization in any case.

Such realities were no doubt also on the mind of V. F. Kliuchnikov, a more moderate commentator, who noted in the same journal in 1911 that realization of "the norms of any kind of forest preservation law at all" could not happen without "an incursion [by the state] into the sphere of individual's and societies' private property rights."[118] This was not stated as part of a case against big-government conservation, but simply to suggest that conservation measures of necessity incurred costs in political and economic freedom. Though this might well be worth paying in some cases, it might not in others, he noted. Certainly, careful thought should be given to the matter.[119] Kliuchnikov, it should be noted, as a typical moderate and utilitarian conservationist nonetheless unequivocally supported state expropriation of forests lands from private owners who did not manage their property in proper accordance with legally-established economic plans.[120]

But it was with the advocates of total state ownership that the future lay, and with them also increasingly the momentum during the First World War and the era of the Provisional Government. During the spring and summer of 1917 in

---

[117] Ibid., 1102-03.

[118] V. F. Kliuchnikov, "Konkretnye zadachi lesokhraneniia," 1410.

[119] Ibid. Like many of his contemporaries, Kliuchnikov tended to favor protections that had clear economic or political benefits, such as those that prevented erosion of navigable rivers, ensured timber supplies, and so on.

[120] Ibid., 1412. Kliuchnikov made this point a second time later in the same essay: "[I]n circumstances where there are problems carrying out [worthwhile forest conservation] . . . under conditions of private property, the state must not refrain from compulsory forest expropriation, and forest conservation law must provide for this" (1424).

particular, as the collapse of the tsarist regime seemed to remove any need for half-measures and incremental change, talk was of little else but outright nationalization. The long-time editor of *Lesnoi zhurnal,* G. F. Morozov,[121] speaking in September at the Second Congress of the Foresters' Union asserted, for example:

> The forest must belong to the state and the latter must be the forest's boss. . . . [T]he lessons and facts of history prove the right of the state to run the forest economy. The state [*gosudarstvennost'*] is the generality of interests; the forest, belonging to the state, belongs thus to all, and only the state can expediently manage it in the interests of all the people. Our task therefore is to strengthen in the consciousness of the people this most important principle, this first . . . axiom.[122]

A clear indication of this same forum's general support for outright nationalization came following the reading of key papers by two respected forest authorities and Union members: D. I. Tovstoles's "On Nationalizing the Forests," and N. A. Kuznetsov's[123] "State forests and their use in connection with agrarian reform." During debate on these papers (both given on 27 September) a motion was approved that the articles, and specifically their call for nationalization, were

---

[121] Along with Vysotskii, Morozov (1867-1920) was one of the most important figures in academic forestry during the late Imperial and early Soviet periods. Much of his research focused on the methods, consequences, and ecology of forest cutting, clearing, and replanting. Indeed, he has been dubbed the "Father of Russian . . . forest science." Morozov dropped out of a military career as a young man to pursue interests in populism and peasant politics. It was via this avenue that he came to study forests, first as peasant environment, then, and for the rest of his life, as biological community. See Ivan M. Elchibegoff, *Georgii Fedorovich Morozov: Father of the Russian Forestry Science* (Washington: Self published, 1967). Elchibegoff notes the apparent incongruity of Morozov's receipt of financial and other professional support from the tsarist government, despite his interest in left politics, but offers no further insights into the man's political mind (7). Morozov's prodigious but short career included 315 "works," chairing the St. Petersburg Forest Society, copious experimental work, many lectures, and editorship of *Lesnoi zhurnal* from 1901 (or 1904 in one account) to 1918.

[122] Quoted in *Lesnoe Khoziaistvo SSSR za 50 let, 1917-1967*, 15.

[123] Kuznetsov (1864-1932) held important positions in the central forest apparatus before and after the Revolution and was a regular writer for *Lesa Respublika* (a Bolshevik forest journal) during 1918-1919. He was one of the main consultants working with N. I. Faleev on the Bolshevik law "On Forests" (see chapter 4). His academic work focused on geobotany in the Caucasus region.

so important that they should be printed up and distributed as widely as possible. Morozov suggested there and then that the fund already existing in his name would provide the necessary financial support for "popularizing the idea of nationalization."[124] The two papers were discussed again on 4 October when their main points, and specifically the need for general forest nationalization, were accepted as a general position and resolution of the whole Congress "without any alterations."[125] The essays themselves were subsequently published in *Lesnoi zhurnal*.

Notions of a foresters' consensus, while useful and accurate to a point, should not be taken too far, of course. In addition to the wide variety of interpretations already noted on the theme of a bigger government for a better forestry, other divisions and complications existed also. (Some of these became especially apparent following the Bolshevik seizure of power, and will thus be discussed in later chapters.) Among the losing minority in voting held in 1911 (in St. Petersburg) on expanding the jurisdiction of 1888's "Statute on the Safeguarding of Forests," for example, were a handful of private owners who were also vociferous opponents of any increase at all in what they saw as government "interference." Far from acquiescing in their peers' general acceptance of bigger government, these called a "special conference" which, after having debated the issue for three days (25-27 January) issued a statement "on behalf of private forest owners" explicitly rejecting the Congress's support of expanding forest protections. Confusingly, however, the following day "a significant number" of these same private owners refused to support the same statement, considering it "not representative" of their own, presumably more moderate, views.[126] Whatever the details, these events certainly highlight both the complexity of assessing forest owners' and foresters' views *en masse*, as well as

---

[124] GARF. Fond 5520, op. 1, del. 2, st. 15.

[125] Ibid., st. 47.

[126] Reported in "Lesnaia sektsiia na pervom kievskom sel'skokhoziaistvennom s"ezde i printsip obshchego lesookhraneniia," *Lesnoi zhurnal* 8 (1913): 1369. The International Congress on Forest Affairs, held in Paris in June 1913 (and which also discussed themes of government regulation and protection of forests), supposedly also expressed a "hesitant attitude" toward such matters. This, however, was in the estimation of one M. S. Eremeev, a private owner of Russian forest land and one of the few outspoken opponents of further government interference.

the doubtfulness of embracing what might seem an intuitive truth: that private owners rallied as one in rejection of any increase of government authority or any diminution of their private property rights. Clearly they did not.

Similarly, even among the advocates of nationalization and centralization there were to be some doubts expressed during the Revolutionary years 1917 and 1918 about the feasibility or wisdom of moving quickly. A report published in the Bolshevik forest journal *Lesa respubliki* (Forests of the Republic) in 1918 (itself fully committed to rapid nationalization), noted rather vaguely that at the Foresters' Union's first delegatory [*delegatskii*] Congress in April 1917 there had been "talk" that it was "still early to go over to nationalization of the forests," that this was "a question for the distant future, [and] that it was necessary to restructure forest life gradually and with great care."[127] In 1918, some Union members also "spoke out against [immediate] centralization so as not to waste time on [further] reorganization."[128]

Still more importantly in considering the limits to any notional foresters' consensus is the thorny question of exactly what was meant when advocating an expansion of state control or ownership. The Russian sources reveal a variety of specific terms, all of which might at first glance appear to have a similar or even identical meaning: *gosudarstvennyi* (adjectival form of "state"); *gosudarstvennost'/gosudarstvo* (variant noun forms of "state"); *narodnyi* ("popular," or "folk"); *obobshchestvlenie* ("socialization"); *sotsializatsiia* (also "socialization,"); *natsionalizatsiia* (nationalization); and still others.[129] In the majority of cases there seems to have been little if any conscious differentiation among these terms, at least prior to October 1917. However, these words, and the concepts they were supposed to communicate, could in fact be seen in at least two distinct ways: as vesting control in the central government or in local populations

---

[127] Ne-Lesovodov, "Lesnaia revoliutsiia," *Lesa respubliki* 16 (1918): 850. Attributed to Professor Khodskii, a member of the Foresters' Union. "Ne-Lesovodov" translates as "not-a-[credentialed] forester." The name—and the similar one, "Ne-Lesovod"—crop up frequently in *Lesa respubliki* during 1918. It appears to be symptomatic of the hostility sometimes displayed by revolutionary forest writers toward traditional, credentialed "experts."

[128] GARF. Fond 5520, op. 3, del. 1, st. 36.

[129] Though translated alike as "socialization," *obobshchestvlennie* has a more traditional, rural, and peasant-based connotation than the loan-words *sotsializatsia* and *natsionalizatsiia*, both of which sound more urban, modern, and strictly "socialist" in the sense that word has in the English language.

and local organs; in other words, as centralizing or decentralizing.[130] During, and especially after, 1917, these two trends came to be more clearly defined as opposites, and more care was used in choosing words—or appropriately defining them—so as to imply clearly either pro-peasant decentralization or more state-oriented centralization.

On the other hand, matters were somewhat simplified by most foresters apparent agreement that forest land differed importantly from agricultural land -- and that it thus neither could nor should be managed or owned in quite the same way. The forests' growing cycles were generational rather than seasonal; their expanse was national rather than local.[131] Based on a shared agreement about these points, a rough consensus appears to have existed that forests, unlike farms, required management as a single entity, on a national scale, and with the long-term perspective of the forest scientist. Above all, they should not simply be transferred to local peasant ownership and exploitation as might be desirable with agricultural land. "The forests are our national inheritance" became a virtual mantra (with the emphasis on "national"). This led naturally enough to talk of central plans and central control. At the same time, however, numerous criticisms were made of the existing Forest Department, even in some cases by its personnel, as "aloof" and out-of-touch with local sentiment and conditions. The rights of the peasantry to adequate timber resources, and of local technicians and planners to due respect and attention, it was claimed, were routinely violated. Less, not more, central control was needed here. These essentially decentralizing sentiments met, however, with their more centralizing counterparts in a roughly common vision of a new forestry whose essential priorities, practices and administration would come from the center, but which would allow expanded local initiative and seek better to satisfy local needs. In short, what was most often mooted was a kind of "organic" or locally-responsive centralism characterized by common work and a shared vision. That this was intended even during the 1900s and 1910s ultimately to translate to central, rather than local

[130] N. I. Faleev noted in this regard that "each person attaches his own interpretations to the concept 'nationalization,' which cannot provide the basis for legislative work." N. I. Faleev, "Kapitalisticheskye osnovy lesnogo khoziaistva," 3.

[131] The First Kiev Agricultural Congress (1913) made such an argument, for example. "Lesnaia sektsia na pervom kievskom sel'sko-khoziaistvennom s"ezde," *Lesnoi zhurnal* 9-10 (1913): 1426-28 and *passim.*

executive and overall control, however, is exampled time and again. The aforementioned Forest Section of the First Kiev Agricultural Congress held in 1913, for example, raised the desirability of various pro-local initiatives. These included: expanding the participation of *zemstvos* in organizing cooperative efforts with forest owners toward "correct organization and exploitation of the forests and regarding their restoration and care . . . reforesting . . . [and] in the battle against erosion";[132] the desirability of establishing within or around local *zemstvos* "bureaus on forestry, forest organization and management," or other *zemstvo*-based forest organs, particularly at provinicial [*gubernskyi*] or district [*uezdnyi*] levels with many or large forests under private or communal ownership; and so on.[133] But the Congress's summary resolution—to create "as many local forest organs as possible"—ended unequivocally with a demand to "unite their actions under a single central organ."[134] Likewise, the resolution in favor of nationalization adopted in September 1917 by the Agrarian Section of the Second Foresters' Union Congress went on to state that:

> of all forms of ownership and use of the forest the single most expedient one [,] and the one best guaranteeing the well-being of the general population of the country—is nationalization of all the forests *under a Central Administration* [Emphasis added]. . . . Nationalization of the forests under a Central Administration allows the participation of the local population and local organs of self-administration, both in the distribution of products of the forest economy among the population and the processing of forest materials, designated for release according to an economic plan.[135]

In just such a way, in fact, had the differing visions of Orlov and Vysotskii back in 1905 been rendered compatible by participants in that congress.[136] Looking forward, and by way of stressing again the continuity in forestry either side of the Revolution, early Soviet-era assertions that Bolshevik forest centralization did not represent "an autocrat" "waving the conductor's wand," but a regime whose entire essence rested on reciprocal, definite, necessary collaboration with the organs of local soviet forest power" seem entirely in accord with this general

[132] Ibid., 1419.

[133] Ibid., 1419-20

[134] Ibid., 1421.

[135] GARF. Fond 5520, op. 1. del 1, st. 61. Emphasis added.

[136] "Protokol' 9-go ocherednogo zasedaniia S. Peterburgskogo lesnogo obshchestva, 26 November 1905," 191, 203, and passim.

impulse.[137] As with other forestry reform plans, however, any hopes for bringing about such a system were ruined at least for the duration, by the outbreak of war in 1914.

Along with general calls for increased state ownership or control, and in close connection with ambitions regarding the creation and management of a single forest resource, most foresters seem also to have agreed on the desirability of a greatly more scientific or "rational" mode of forest management. In large part, these urges were inseparable from the wider goals and plans already outlined. In fact, science had already begun to play a significant role in the forest economy as evidenced, for example, in the earlier-noted work on forest-related entomology. Pre-Revolutionary Russia also made important progress in the creation of forestry schools and in the training of specialists—this long before such education was available in the USA[138]—and otherwise in co-pioneering with other nations experimental scientific forestry.[139] In this regard, during the 1900s and 1910s numerous experimental forests (*opytnye lesnichestva*) were set up, in

---

[137] N. I. Faleev, T. Nekhoroshev, D. Sazhin, "Lesnye organy," *Lesa respubliki* 5-7 (1918): 340. See also the "General Rules" (articles 1-4) of the Basic Law on Forests, which make a similar point.

[138] The education of forestry specialists dates in Russia from 1803, with the organization in Pushkin (formerly Tsarskoe Selo) of a forest *uchilishcha*. Here, twenty students undertook a three-year course of study. The country's first formal Forestry Institute was opened in 1805 and also offered a three-year course of training. In 1808 in St. Petersburg, a *uchebnoe zavedenie* was opened. It too was called the Forest Institute. As the Petersburg Forest Institute, this became the major center for forestry studies, absorbing the two earlier schools in 1811 and 1813, respectively. By 1917 it had prepared about four thousand specialists. In 1928 it was renamed the Leningrad Forest-technical [*lesotekhnicheskaia*] Academy (LTA). The training of forest specialists in tsarist Russia was carried out also at the *Petrovsko-Razumovskaia zemledel'cheskaia i lesnaia akademiia*, founded in 1865. Beginning in 1869 forest specialists were trained also at the *lesnoi fakultet Novo-Aleksandriiskogo instituta sel'skogo khoziaistva i lesovodstva* near Warsaw and at the *Gory-Goretkyi sel'skokhoziaistvennyi institut* in Belorussia. A number of lower [*nizshie*] schools for the training of less skilled forest technicians were also opened. These numbered 10 in 1888, 32 in 1907, and 43 by 1913. Graduates typically worked as *konduktury*, though in some cases they continued their studies at the Petersburg Forest Institute. See *Lesnoe Khoziaistvo SSSR za 50 let, 1917-1967*, 271-72.

[139] Russia's first experimental work in forestry is associated with the name Victor Egorovich Graff. For more on Graff and forest science in general, see, ibid., 85-86.

most cases under the auspices of the St. Petersburg Forest Society, the Forest Department, or both, and located near Russia's two biggest cities.[140] A perusal of scientific forestry experiments either already underway or to be carried out in these forests—noted at the 11th All-Russian Congress of Forest-Owners and Managers held in Tula in 1909, for example—gives an idea of the momentum already existing behind such work.[141] Also part of the pro-science and planning consensus was the widely-expressed desire for more and better-trained forest specialists. A presenter from the Forest Section of the 1913 First Kiev Agricultural Congress, for example, considered it imperative individuals with the "highest forest education" play a key role in forest management, particularly in privately-owned forests.[142] Current legislation paid lip-service to the deployment of such cadres it was noted, but provided little of real substance.

Education of the public on matters of forest conservation was also deemed important. There was little dispute about either the nature or the difficulty of the basic challenges involved. It was widely acknowledged, for example, that most peasants' understanding of the forests rested not on reason but on superstition and tradition. Peasants tended to see the forest as a "gift from God" to their communities. Forests were not the result of human labor, nor could they improved by such. The forest, peasants insisted, "was everyone's and could be used by anyone who wished."[143] The notion that portions should be locked up or the whole managed according to the "scientific" principles of "outsiders" was very dimly viewed. Such opinions, foresters believed, must be changed.

---

[140] Though these were relatively busy years in the establishment of experimental forests, they were by no means the beginning. G. F. Morozov had been involved in the creation of several dozen "experimental forest strips" in the kamenno-stepnoi experimental forest district (probably in Voronezh) during the 1890s, as well in state-sponsored work reafforsting areas characterized by sandy soils in experimental forests in Pavlovskii county, Voronezh province in 1898. Even before these, at least three artificial steppe forest districts had been created by the illustrious Russian soil scientist V. V. Dokuchaev. See, Ivan M. Elchibegoff, *Georgii Fedorovich Morozov,* 7.

[141] "Rezoliutsii XI vserossiiskogo s"ezda lesovladel'stev i lesokhoziaev v g. Tule," *Lesnoi zhurnal* 1-2 (1910): 4-10.

[142] "Lesnaia sektsiia na pervom kievskom sel'skokhoziaistvennom s"ezde i printsip obshchego lesookhraneniia, chast' vtoraia," *Lesnoi zhurnal*, 9-10 (1913): 1418.

[143] Ibid., 1433. A similar estimation of popular views of the forest is offered in "Beseda o lese i ego znachenie v zhizni naroda," 7-8.

"Education" was the envisioned cure. There was, however, little clarity about the precise manner in which the transformation would be effected. Pamphlets (for the minority of peasants who could read) better links between central and local forest organs, verbal propagandising, and so on were the usual ideas. The prospects for success were viewed variously. A 1916 congress of foresters of the 3rd district of Tambov *gubernia*, for instance, agreed on the possibilities for success of a two-pronged approach that would simultaneously convince the masses of the importance of forests as an "active factor" in the "natural-historical environment" of every region and at the same time as an important component in the "spiritual and material life of humanity."[144] A less optimistic, but—it would turn out—probably more realistic commentator was an anonymous speaker at the All-Russian Extraordinary Congress of Forest Owners and Managers, held in January 1911 in St. Petersburg, who spoke of the need to reckon with the reality of peasant non-cooperation in matters of forest organization and conservation. He urged instead (though ambiguously) that "another path toward forest conservation" be found.[145]

Among the various more minor points or concerns raised by foresters during these years, brief mention should be made of criticisms of the system known as "servituty," which was singled out on several occasions as a cause of unacceptable forest destruction. Defined typically as "legal limitations on private property rights," *servituty* (the word is plural in Russian) existed for the most part only in western provinces, including Kiev, Podol'sk, and a few others, as well as Poland and the Baltics. *Servituty* arose as one of the many variations on land-holding and land-use rights after the abolition of serfdom in 1861. In the manifestation under review here, *servituty* was designed to help peasants who held insufficient land by granting them the use of specified resources on land held privately by other persons or groups, or by the state.[146] Common benefits included the right to drive or graze cattle and to collect forest materials. It was precisely these activities, and their relatively unregulated nature, which attracted criticism. Speakers at the 1909 Tula meeting noted earlier deemed the practice

---

[144] "Zhurnal S"ezda Lesnichikh III-ego raiona Tambovskoi gubernii v 1916 godu," *Lesnaia zhizn' i khoziaistvo* 2 (1917): 19.

[145] "Lesnaia sektsiia na pervom Kievskom Sel'skokhoziaestvennom S"ezde i printsip obshchego lesookhraneniia," *Lesnoi zhurnal* 8 (1913): 1368.

[146] *Pol'naia entsiklopediia russkogo sel'skogo khoziaistva* (St. Petersburg, 1903).

generally "harmful to the preservation of private and state forests and causing extreme complications in the carrying out in them of a proper economy," and on this basis promoted its reform.[147]

Goals of Forest Protection

Such then was the relatively advanced, and certainly lively, condition of Russian forestry—its practice and theory, and the activities of the forestry community—on the eve of the revolutions of 1917. The foregoing discussion has focused almost entirely on the *methods* of better forestry. Before moving on to the events surrounding Russia's political convulsions during 1917, it will be necessary to add a few words assessing the *reasons* or *purposes* for which Russians felt better management and protection was desirable or necessary. In fact, far less attention seems to have been given to this question by foresters themselves, the answers, perhaps, having been considered obvious, or at least less subject to dispute. Here and there, however, such matters were treated explicitly.

For most of the nineteenth century, and in approximate order of priority, the stated goals of forest protection appear to have been: (1) assurance of vital timber and wood fuel supplies for the needs of state (including export) and society, now and in the future[148]; (2) prevention or minimization of soil erosion and inundation, and of the various problems associated therewith, especially threats to navigation, timber floating, etc.; and (3) climate stability.[149] During the

---

[147] "Rezoliutsii XI vserossiiskogo s"ezda lesovladel'stev i lesokhoziaev v g. Tule," 10. For more information on forest conservation and *servituty*, see: N. I. Faleev, "Otmena lesnoi sobstvennosti." In N. I. Faleev, ed., *Osnovnoi zakon o lesakh, izdanie neofitsial'noe* (Moscow: Izdanie Narkomzema, 1918), 62-66. For general information on *servituty*, see Abramovich, *O krestianskykh servitutakh* (1895) and Novitskyi, *Servituty i obiazatel'nye razverstanye v iugo-zapadnom krae* (1881).

[148] As early as the mid-nineteenth century it had been noted that more efficient exploitation of Russia's forests could yield income equal to half of the total state budget at that time. See G. Piatakov, "Kontsessionaia politika v lesnom dele," *Lesopromyshlennoe delo* 5-6 (1923): 3. Similar or comparable opinions held up to and even beyond 1917.

[149] See for example, V. F. Kliuchnikov, "Konkretnye zadachi lesookhraneniia," 1410-24. Kliuchnikov and others also sometimes took issue with what they considered excessively dramatic and unscientifically-based arguments especially about the vital importance of forests for climate control. As

last two or three decades prior to the First World War, however, other, less tangible considerations began to assume importance. These impulses—scientific, moral, and aesthetic—and the cohort of individuals promoting them lie at the center of an account of early Soviet nature conservation told by a variety of writers.[150] In it, a growing sense of the fragility of nature, evidenced by obvious physical changes in the Russian landscape, and the influence of a growing international conservation movement—especially the pioneering and proselytic efforts of the Prussian conservationist Hugo Conwentz—provide early tributaries to the swelling flood of nature conservation advocacy in Russia itself, building on early, tentative steps taken by groups such as the Moscow Society of Nature-Lovers (*Moskovskoe obshchestvo liubitel'ei prirody*) and others. A critical major step forward occurs in 1908, when Moscow zoologist G. A. Kozhevnikov outlines proposals for a network of *zapovedniki* at a Jubilee Convention of the Imperial Russian Society for the Acclimatization of Animals and Plants. The blueprint was publicized at least twice more over the next two years, by Kozhevnikov himself and by I. P. Borodin, before emerging into the light of day as the Soviet network of *zapovedniki* refered to in the introduction.[151]

Though always told as part of the history of conservation, never of forestry *per se*, in fact (and obviously) the forestry community had close links with all these developments, as well as overlaps with its personnel. Though forest scientists were almost certainly among Kozhevnikov's original audience in 1908, most, perhaps, heard about the project from P. Vinogradov-Nikitin, another forest expert who, having heard a more detailed presentation by I. P. Borodin on 22 December, 1909 at the Imperial Geographical Society, published a highly favorable report in the January-February issue of *Lesnoi zhurnal.*[152] Borodin himself included forests—in general and certain specific threatened ones—among his priorities for areas worth designating "monuments of nature." And N. I.

---

earlier noted, debate on these issues bears considerable resemblance to recent disagreements about the reality or perniciousness of global warming (1420-21).

[150] See in particular, Douglas R. Weiner, *Models of Nature,* chapter 2.

[151] A brief contemporary overview of these events is given by a participant, A. P. Semenov-Tian-Shanskii, in "Svobodnaia priroda, kak velikii zhivoi muzei, trebuet neotlozhnykh mer ograzhdeniia," *Priroda* 4-6 (1919): 200-16.

[152] P.Vinogradov-Nikitin, "Okhrana pamiatnikov prirody," *Lesnoi zhurnal*, 1-2 (1910): 146-54.

Kuznetsov, a key figure in late Imperial and early Soviet forestry who was also present at the lecture, actively debated Borodin, whose ideas he supported.[153]

Similarly, and as already discussed, foresters and forest scientists also continued through the immediate pre-Revolutionary period to play a key role in putting the study of natural communities, in this case forests, on a more scientific footing, thus helping shape the agenda and form of early Soviet conservation in general and the *zapovednik* project in particular. Overall, it can fairly be stated that tsarist-era forestry included sufficient important strides made in the scientific study of forests to render the Bolsheviks' preoccupation with forest science and planning (see chapters 3-7) an expansion of pre-Revolutionary trends rather than a novel departure. (The same, incidentally, may be said regarding forest protection based on moral and aesthetic values and priorities, which was promoted by Borodin, A. P. Semenov-tian-shanskii, and others, and which Soviet legislation was also to pay at least lip service to [see chapter 5]).

* * *

By 1914 then, and even more noticeably by early 1917, the various strands of Russia's forest protection movement—and of the wider trend around it toward more efficient and well-organized resource management overall—had increasingly come together, creating a relatively unified movement possessed of a set of generally-agreed-upon ideas and goals. These reduced essentially to the establishment of a single, rationally planned national forest economy, utilizing scientific methodologies and technical knowledge, managed in consensual association between center and localities (though with the center clearly in ultimate control), and based on state ownership (however construed). Whether or not the momentum thus established would have been sufficient to bring about real reform in the absence of major political upheaval is difficult to judge. Probably, the tsarist government, a bastion of conservatism, and the wider social structure around it based essentially on private property would have proved insurmountable obstacles. On the other hand, it seems at least conceivable also that Russia's considerable tradition of reliance on the state—most noticeable in its industrialization—and even its quasi-socialistic forms of peasant association in the countryside might have provided both precedent and basis for eventual implementation of some of the "big government" conservation schemes already

[153] V. F. Kliuchnikov, "Konkretnye zadachi lesookhraneniia," 1416-19.

examined. In February 1917, however, the tsarist regime was swept away forever; and Russia, its forests as everything else, headed into uncharted territory.

III
Plans and Realities:
Forest Organization and Protection from February to October 1917

The collapse of the tsarist regime in February 1917 (and the subsequent establishment of the Provisional Government[1]) had two immediate consequences for forest affairs. In the first instance, the sweeping away of established authority opened wide the possibilities for major and rapid reform. Overnight, plans and programs worked out during the preceding years based on nationalization, scientific methods, and central planning were transformed from impossible dreams or very long-term projects into tantalizing potential achievements. As one forest worker put it, the approaching spring season had brought with it also a "political spring." The feeling was widely shared; and similar metaphors were suddenly in common use in forestry circles. The season became "radiant," "millennial," and so on. One individual spoke of a "general fever"[2] seizing hold of large numbers of foresters as they came together that April from all across the country to Petrograd to evaluate the new environment at the First Congress[3] of a new umbrella organization called the Foresters' Union (*Soiuz lesovodov i lesnykh*

[1] Following the rapid and unanticipated collapse of tsarist authority a new, temporary government was cobbled together from the Russian parliament. Known as the Provisional Government it was dominated by relatively liberal, westernizing elements. From April forward, Lenin and the Bolsheviks criticized it mercilessly, urged Russians to reject it, and actively plotted its overthrow. They succeeded in October 1917.

[2] GARF. Fond 5520, op. 2, del. 16, st. 88.

[3] A smaller meeting of a rump of founding members took place a little earlier, on 26 March. To make matters still more confusing, the "First" Congress noted here was followed in the fall by a second one entitled the "First Delegates' (*delegatskii*) Congress. The "Second Delegates'" Congress, held the following spring was, accordingly, the third major congress overall. See, for example, "Vtoroi delegatskii s"ezd Soiuza lesovodov," *Lesnoi zhurnal* 3-5 (1918): 336. For more on the general mood of the gathering see *Lesa respubliki* 16 (1918): 849. On the founding and goals of the Foresters' Union in general, see either "Soiuz lesovodov i lesnykh tekhnikov," *Les* 5 (1917): 24-25; or "Vserossisskii Soiuz Lesovodov i Lesnykh Tekhnikov (Konduktorov)," *Lesnaia zhizn i khoziaistvo* 2 (1917): 45.

rabotnikov, or *Soles* for short).[4] Giddy optimism aside, perhaps the most immediate tangible advantage of the new order was precisely the greater ease with which such forestry societies and organizations could now be created and their agendas written or changed. While the Foresters' Union was the most obvious beneficiary, smaller groups also found cause to rejoice. The minutes of the "New Forest Society"—a small Ural-region outfit founded in Ekaterinburg on 5 March—noted, for example, that such activities would now "no longer provoke the huge amount of fuss and complications that would have been the case earlier, that is, before the Russian people's achievement of freedom of association and other freedoms."[5]

On the other hand, however, early optimism was very soon diminished by the actions of another constituency, one possessed of its own agenda for Russia's forests and in no mood to wait for directions from the Forest Department, major forest congresses, or any other authority. Rural populations—encouraged by the sudden evaporation of political and legal authority, in many cases disillusioned and hardened by the experiences of war, and generally hungry for forest materials—descended that spring in their thousands on Russia's forests, beginning a devastating period of plunder and cutting all across the country[6]. A near-contemporary assessment cast these developments in the bleakest of lights:

> In the localities, in forest affairs, the revolution arrived in February. Here and there the population . . . felt the beginning of a revolution and, having felt it, related to it with a complete lack of awareness of what was going on, with complete disdain for the future. . . . The slogan 'freedom', during March . . . was taken by the population as permission to do the impermissible, as the opportunity to bring about what had before been unachievable. The forest [they said] is no-one's, it is God's, the forest is not cultivated,

---

[4] The Union was from the start a politicized body. It explicitly supported the Provisional Government. Partly for this reason it was from the start viewed with great suspicion by the Bolsheviks. On the Union's political ideals and associations, see *Les* 5 (1917): 24.

[5] "Novoe Lesnoe Obshchestvo," *Les* 5 (1917): 11. The same source carries on to relate the society's prehistory under less favorable circumstances in the region going back to 1910.

[6] There were, however, important regional variations, with many of the worst excesses being reported in south-central provinces, such as Tambov, Voronezh, and so on. Here, the relatively high proportion of persons to forests made timber an especially valuable commodity, as well as increased the pressure for conversion of forest land to agriculture. Forest-rich and less populated areas in the North and beyond the Urals, while they suffered varying damage also, were hugely better able to withstand the impact.

> not planted, the forest is [simply] 'ours'. The pressure on the forests assumed wild forms. Everyone who was able or so desired cut down the forest. They dealt with the forest as with an eternal enemy. It seemed as if the *revolution* was being directed specifically against *the forest* rather than *against the forest regime* [emphases in original]. The excesses were horrifying. The cutting of forests assumed an insane character. Everything at hand was cut down. Every effort to stop the anarchy in forests matters was doomed beforehand to failure.[7]

It was this depressing reality, far more than any hypothetical bright future, which began increasingly to command the attention of forest specialists as the year progressed. So desperate did the situation appear by the end of the year that in opening the Moscow Regional [*oblastnoi*] Congress of the Foresters' Union in January 1918, the chair, V. I. Stankevich, called for the slogan adopted by the All-Russian Congress in April—"Take Care of the Forests"—to be changed "to a new more alarmist" one: "Save the Forests from Anarchy and Destruction!"[8]
The wrath of Russia's peasants was not confined only to the forests themselves. The existing corpus of forest specialists and guards—essentially the very individuals and groups aspiring to shape and lead Russia's new forest order—themselves began to suffer frequent attack. The following, for example, was reported in Tambov province shortly after the February Revolution:

> In a house at Fashchevskoe *lesnichestvo* at the beginning of March [1917] during a meeting of local intellectuals regarding the organization of local [*volostnye*] and rural [*sel'skye*] executive committees a crowd of drunken peasants burst in, headed by the former clerk of the forester. They dispersed the meeting, arrested the elderly forester and then began to accuse the members of the meeting of faithfulness to the old regime. The . . . forester himself was so affected by all this that he [subsequently] resigned.[9]

Though scenes such as these were common enough in many parts of Russia, Tambov actually appears to have seen more than its fair share of forest-related disturbances. Some of the most intense unauthorized cutting also appears

---

[7] Ne-Lesovodov, "Lesnaia revoliutsiia," *Lesa respubliki* 16 (1918): 848.

[8] V. M. Bortkevich, "Moskovskii oblastnoi s'ezd Soiuza lesovodov," *Lesnoi zhurnal* 3-5 (1918): 113.

[9] "Polozhenie lesnogo dela v Tambovskoi gubernii s nachala revoliutsii po 15 maia n. s. 1918 goda." GARF. Fond 5520, op. 2, del. 16, st. 88. This is an anonymous essay filed in the archive of the All-Russian Union of Foresters.

to have taken place here, too.[10] In part, this is likely a function of basic economic and demographic facts: Tambov and neighboring provinces were relatively heavily populated but lightly endowed with forest—making timber especially scarce, and questions of access and control particularly troublesome. Population pressures also led to continual conversion of land to agriculture, further reducing the forests.[11] Tambov is of further interest, however, since during 1920-1921 it was to become the scene of a major and more general anti-Soviet peasant uprising—the so-called "Antonovshchina," named after its leader, Antonov. By the summer of 1921, the Soviet authorities' efforts to crack down on Tambov-area unrest had resulted in the incarceration in concentration camps of some 50,000 "bandits" and "members of the families of bandits taken as hostages."[12] "Large-scale executions of hostages and the bombing of villages with poison gas" were also among the techniques employed here and elsewhere as the Bolsheviks strove toward the "eradication of bandits from the forests."[13] Though the Antonovshchina revolt in particular—and Tambov-area rural uprisings in general—are more typically associated by historians with resentment over state grain-procurement efforts it seems likely that forest-related tensions played a role too.[14]

By late spring 1917, scenes such as that involving the "elderly forester" noted above—and some far worse—had already become common. Throughout the year and beyond reports flooded into local and central branches of the Foresters' Union detailing and condemning forced evictions of forest personnel from their assigned *lesnichestva*, and many other abuses. The victims—ranging

---

[10] I have not been able to substantiate this claim with hard figures. It is based only on the relatively high frequency within the relevant source materials of references to specific events or general trends in and around Tambov. The great historian of Russia W. H. Chamberlin, however, noted that more than half of all agrarian disturbances registered in 1917 took place in the Central-Agricultural and Middle Volga regions, notably in Tambov, but also in Voronezh, Orel, Kursk, Tula, Riazan, and other nearby regions. Chamberlin, as cited in Sir John Maynard, *The Russian Peasant and Other Studies* (New York: Collier, 1942), 91.

[11] See for example "Doklad Raionnomu S"ezdu Lesnichikh IV raiona tambovskoi gubernii," *Lesnaia zhizn' i khoziaistvo* 1 (1917): 9.

[12] Stephane Courtois, et al., *The Black Book of Communism: Crimes, Terror, Repression* (Cambridge, MA: Harvard University Press, 1999), 80.

[13] Ibid., 109, 116-17.

[14] On the Antonov revolt see, Oliver H. Radkey, *The Unknown Civil War in Soviet Russia: A Study of the Green Movement in the Tambov Region, 1920-1921* (Stanford, CA: Hoover Institution Press, 1976).

from learned foresters to simple forest guards—were in many cases robbed and beaten. Many lost their places of residence. Death threats were made. Occasionally, forest guards were in fact killed, as in one reported case in the *Stolovoe lesnichestvo*, also in Tambov province.[15] At the same time loud criticisms were voiced of the current head of the Forest Department, N. Grudistov,[16] of the "sclerotic and corrupt" center in general, and of "outsiders" overall. As summer wore on and the Russian war effort collapsed, further and further waves of peasants flowed back from the front into the villages and forests, compounding the problem. Whether animated by revolutionary ideals or simpler hatreds, returning peasant soldiers seemed in little mood to accept restrictions on their activities in the forest. One observer spoke of a "bad scent" that began to fill the air as everywhere rural populations rose up against their former forest administrators, branded them "people of the old regime" and "enemies of the people," and pushed them out of the *lesnichestva* altogether.[17]

Even given such circumstances, optimism remained, however, that foresters and peasants might still see eye to eye regarding a post-tsarist forest order. There were, after all, important core areas of agreement between the two constituencies. Both peasant and forester concurred, for example, that the old forest apparatus had been excessively "aloof" in its dealings with provincial populations and priorities, and that it must, therefore, adopt attitudes and policies more congenial to these. Both shared also a sense of relief at the collapse of Russia's traditional order and hoped for major changes in the future. More concretely, and above all else, there appeared also to be a common yearning for nationalization—foresters and peasants alike regularly spoke of the forests as a

[15] "Polozhenie lesnogo dela v Tambovskoi gubernii s nachala revoliutsii po 15 maia n. s. 1918 goda." GARF. Fond 5520, op. 2, del. 16, st.st. 88-89.

[16] Ne-Lesovodov, "Lesnaia revoliutsiia," 849. Grudistov was replaced during 1917 by N. I. Faleev, who will feature significantly in later chapters.

[17] "Polozhenie lesnogo dela v Tambovskoi gubernii s nachala revoliutsii po 15 maia n. s. 1918 goda." GARF. Fond 5520, op. 2, del. 16, st.st. 88-90. In many cases, forest guards not driven out left anyway. Increasing work loads, poor and erratic pay, the unpleasant and sometimes dangerous circumstances in the forest all contributed. Forest workers were also called up for military service in large numbers throughout the year and during the ensuing Civil War period. Others, including several prominent specialists, emigrated to the Ukraine (see chapter 6).

"the inheritance of all the people" (*vsenarodnoe dostoianie*) for example—and it was here in particular that many foresters pinned their hopes for future reform and improvement even as the forests cracked and fell around their heads. Some cited a peasant *nakaz* (document containing a set of instructions or opinions) compiled during 1917 by the editors of a newspaper, the "News of the All-Russian Soviet of Peasants' Deputies" (*Izvestia Vserossiisskogo Soveta Krestianskykh Deputatov*) on the basis of 242 smaller local peasant *nakaz*es.[18] This "recognized the need to eliminate 'private' property in land and forests," as a matter of basic "justice,"[19] and echoed in this regard dozens of other similar peasant statements before and after. In this respect, of course, it also accorded with forester opinion on the same issue. Participants of the Foresters' Union's spring 1917 Congress also heard that there was "support" among rural populations for their "idea of nationalizing the forests—in the sense of designating them all state [*gosudarstvennyi*] property and exploiting them in the interests of the population with the [latter's] direct participation."[20] And in May of the same year, the Union's Forest-Political Section considered "reports [given by] local delegates regarding the attitude of the peasant population to forests and to the forest administration" which noted the "existence of widespread opinion about the necessity of transferring the forests either to [peasant] societies [*obshchestva*] or to the government," and the need to grant to the population in general "the right to manage or control the forest economy itself."[21] Not surprisingly, the Section concluded that "the idea of nationalization of the forests in the sense of accepting all forests as the property of the state [,] and the exploitation of them in the interests of the population by means of its direct participation therein has support in the minds of the people in most parts of the state."[22] It seemed, in other words, that the "foresters' consensus" discussed in chapter 2 might be shared also by the general rural public.

---

[18] The *nakaz* was subsequently used as a basis for discussion on land reform at the Constituent Assembly (decreed March 1917, dispersed by the Bolsheviks January 1918). For more on the Constituent Assembly, including peasant politics therein, see Oliver H. Radkey, *The Election to the Russian Constituent Assembly of 1917* (Cambridge, MA: Harvard University Press, 1950).

[19] See N. Zubovskii, "Tsentrovye s. r. i lesa," *Lesa respubliki* 5-7 (1918): 292-93.

[20] GARF. Fond 5520, op. 1, del. 1, st. 54.

[21] Ibid., st.st. 54-55.

[22] Ibid., st..55.

In fact, this was not so. The ambiguity of terms such as "nationalization," "popular" and "state"—which could imply either central or local, planned or spontaneous use—has already been noted. And the same is true of the term "property of all the people." For local organs and populations, all such words reduced easily into claims that forests were "the property of a given village or *volost* '"—from which perspective it could thus be argued that "the existing [central] forest administration" was not so much in need of reform as altogether "unnecessary."[23] For foresters, on the other hand, the same words implied centrally-managed planning *on behalf of* the national population or state. Even the above-cited peasant *nakaz* was riven with this kind of ambiguity. It stated, for example, that forests having "general state significance" were to be "transferred to the exclusive *use* of the state" while "all minor [*melkye*] forests" were to go over "to the *use* of communes, where they will be managed by local organs of self-government [italics in quoted source]."[24] Nothing was said at all regarding the *actual proportion* of forest to be declared of "general state significance" to that of "local use," or what constituted a "minor" forest. If it was possible in 1917 for hopeful foresters to imagine they and the rural population might have something similar in mind, it was not to be so for much longer.[25]

Foresters were not blind to these and other challenges. If they were optimists for the long term, many appear to have been less sanguine about immediate prospects. Great attention was devoted throughout the era of the Provisional Government and beyond to evidence of the general antipathy among local populations to forest planning in general and forest protection in particular. "The peasantry wishes to get forest materials free of charge and for whatever

---

[23] "Polozhenie lesnogo dela v Tambovskoi gubernii s nachala revoliutsii po 15 maia n. s. 1918 goda." GARF. Fond 5520, op. 2, del. 16, st.st. 89.

[24] N. Zubovskii, "Tsentrovye s. r. i lesa," 292-93.

[25] 1918's "On Forests," while it allowed for the creation of minor forests and for their exclusion from the law's jurisdiction, placed overwhelming emphasis on the creation of a single, national forest of general state significance, essentially reflecting state and forester conceptions of the split between the two types of forest. On the other hand, 1923's "Forest Code" more closely reflected peasant interpretations of the split, and dramatically redivided the forests, hugely upping the proportion to be excluded from general state holdings. See chapters 5 and 7 in particular.

purpose they wish,"[26] it was noted in 1918 at a session of the All-Russian Soviet Central Executive Committee *(Vse-Rossiiskii tsentral'nyi ispol'nitel'nyi komitet*, hereafter, VTsIK). Some months earlier it had been claimed that provincial populations in general viewed the whole idea of forest preservation "with extreme distrust."[27] This, in turn, essentially repeated the conclusion reached by an agricultural congress held in Kiev in 1913[28]: that the "Russian folk cannot come to terms with the idea of forest conservation [at all]." Behind this resistance lay those seemingly intractable popular notions about the nature of the forest itself that educated foresters seemed destined forever to battle against in vain: that it was a gift from God to the peasant and needed no particular care or restraint; that conservationists were meddling outsiders only concerned with controlling the rural populations for their own, nefarious ends; and perhaps most frustratingly, that "no matter how much [forest] you cut, you still can never cut it all, and even if you did cut it all it would easily grow back by itself."[29] And so on. Notions of conservation may even have run counter to popular peasant notions of personal freedom and domestic order: "Who is not a thief in the forest is not boss at home [*kto v lesu ne vor, tot doma ne khoziain*]," one speaker noted, was a popular peasant axiom and one more obstacle to winning over the peasant.[30]

And there were still other depressing indicators. As late as the early 1900s (and probably through the Revolution) peasants in many parts of the country still regularly set forests alight to drive animals, causing obvious damage. And as already noted, generations of would-be forest fire fighters had wrestled—and continued to grapple—with the peasantry's general lack of interest in helping extinguish fires except when promised cash rewards. There were also innumerable anecdotes in circulation of specific and destructive peasant opposition, such as tale of one individual who had "destroyed an ancient apple orchard [*vekovaia iablonia*] in three days of fear that it might be declared a

---

[26] RSFSR, *Protokoly zasedanie Vserossiiskogo tsentral'nogo ispolnitel'nogo komiteta, 4-go sozyva. Stenograficheskii otchet* (Gosudarstvennoe izdatel'stvo, 1922), 318.

[27] N. I. Faleev, "Chto sdelano i chto predstoit' sdelat'," *Lesa respubliki* 8-9 (1918): 467.

[28] In at least one case, these conclusions derive from the earlier 1911 St. Petersburg Congress and were simply reiterated here.

[29] "Beseda o lese i ego znachenie v zhizni naroda," 5.

[30] Ibid. Various other such examples abound in the literature of the period. For a more general impression see also N. I. Faleev, "Kapitalisticheskye osnovy lesnogo khoziaistva," 6.

[protected] monument of nature."[31]

During 1917 members of the Foresters' Union and the Forest Department hurled particular invective at the activities of local guards set up by peasant groups following the expulsion of those appointed by the old center. These, supposedly, provided "no protection at all," but, operating under what were characterized as typical rural priorities, simply sanctioned the worst abuses in the short-term interests of local populations. The question of how to rectify this particular problem received significant attention at a variety of forums. Beyond exhortations and pleas, however, little could be done at the moment.

In June 1917 the Executive Committee of the Forest Congress and the Council of the Foresters' Union (*Ispolnitel'nyi komitet lesnogo s"ezda i Sovet lesnogo soiuza*) wrote a joint letter to the Ministry of Agriculture (specifically to N. I. Faleev, at that time Assistant Minister of Agriculture, but soon to become chief forester and then the principal architect of the Bolsheviks' first major forest code). The letter expressed great concern over the condition of forests in the Orlov and Kursk regions where the "interference of the population and of their elective organs in forest economy techniques violates the entire plan of forest management." "The taking away of . . . land from forest officials and guards by decision of the Land Committees, with virtually no monetary compensation," the note continued, "threatens to leave these forests entirely unprotected."[32] Information of the type contained in the note cannot, of course, have come as news to Faleev any more than to any other informed person. Nor does Faleev appear at any point to have harbored romantic illusions about the typical peasant's readiness for planned, national, and scientifically-organized forest exploitation and conservation. Writing shortly after the Bolshevik Revolution in his new capacity as chief forester, he noted that "in the forest the peasant sees only the [means of] support[ing] his own needs." Therefore, establishing "correct forest economics" would entail "forcing" him to forego his immediate self-interest for the sake of the national economy, and to leave the forest alone "for many long years" so that it might grow and provide mature timber. Faleev himself did not think the peasantry could or should be thus reoriented in a single,

[31] Cited in P. Vinogradov-Nikitin, "Okhrana pamiatnikov prirody," 149. The original citation is from 1906.

[32] GARF. Fond 5520, op. 1, del. 34, st. 84.

rapid step,[33] but saw the need for great patience and flexibility.

It should not be assumed, however, that among Russia's peasant communities there was no interest at all in any type of forest conservation. Rural populations did sometimes take steps to protect against—though far more often to punish after the fact—unauthorized or destructive cutting. An article published in 1911 in *Lesnoi zhurnal* offers an intriguing vignette:

> The attitude of the peasants toward unauthorized felling is essentially patriarchal: the matter is usually dealt with by having the culprit buy the *mir* [peasant commune] a round of drinks. In the village of Trostianka in Samarskii *uezd* a person caught stealing [timber] was berated [*ego branili*] at a meeting and spat at in the face. [The village then] had a round of drinks at his expense and let him go.[34]

Victims of this and similar rituals, however, were punished not so much for any sins against the forest or the environment, but for having put their own personal needs ahead of those of the village. The well-being of the commune, not of the forest, mattered most to peasants in the estimation of academic and institutional foresters. Thus, there there was little patience for "peasant approaches" of this sort. They were seen as unscientific, ineffective, hopelessly localist and insular in orientation. Moreover, they were impossible to fit into a larger national scheme. Rather than acquiescing in the population's localist priorities, foresters during the era of the Provisional Government and after held that the peasantry, not they, would have to change. To this end "cooperative efforts" would be required to "clarify for the population" the meaning and "great importance" of state forest preservation,[35] to rectify the peasantry's failure to "properly master"[36] the

---

[33] N. I. Faleev, "Otchuzhdenie lesov." In N. I. Faleev, ed., *Osnovnoi zakon o lesakh, izdanie neofitsial'noe* (Moscow:: Izdanie Narkomzema, 1918), 72.

[34] I. Gershman, "Ocherk istorii lesovladeniia," 515. The same source also noted another peasant custom whereby plundered materials carried off by hand were ignored, whereas those taken on horse or other pack-animal were dealt with in the manner already described; and a third whereby violators considered by local peasants to be "wealthy," but not any others, were beaten (ibid.). Several of the various anecdotes presented in this study attesting to peasant attitudes towards forest appear to date from agrarian unrest during 1905-1906.

[35] GARF. Fond 5520, op. 1, del. 1, st. 54.

[36] Ibid., op. 2, del. 16, p. 93.

concepts of "nationalization" and "socialization," and so on.[37]

By mid-1917, then, a conflict was clearly emerging—one long in gestation, but now, in conditions of political collapse, becoming manifest and obvious. It pitted foresters' increasingly consensual general plans for a locally-responsive centralism, managing in scientific fashion a single "national inheritance," against the quite different plans of Russia's rural populations, and of their Socialist Revolutionary representatives. The latter groups gave every evidence of desiring local autonomy, locally-organized forest regimes, and little to no input from scientists, planners or other "experts" or outsiders of any sort. These were the emerging battle lines in a struggle over the shape and dynamics of post-tsarist forestry, itself a microcosm of a wider battle over the future of the whole Russian Empire (see concluding chapter).

In the meantime, however, if foresters were worried, peasants themselves, while they still found much to complain about, must also have had much to rejoice in. Sources of timber which they had earlier been denied altogether or to which they had been provided only limited access were now already on their way to the village. Specialists and "representatives of the Old Regime" were being sent packing in droves. And those still at their posts could now be challenged and ordered about as never before, as is clearly suggested in the following report on affairs in Tambov province during 1917 and early 1918 filed by a member of the Foresters' Union.

> Every delay in fulfilling the demands [of the peasants] has resulted in [their] dissatisfaction, [in] profanities, complaints, has intensified thefts and [other] unauthorized actions. . . . [The peasants] are unhappy with the norms of timber distribution. They refuse building-timbers from deciduous forests, they want pine. They are not happy with their allotted cutting areas, the forest is no good, it's far from their village, it's a swamp. They are not satisfied with the slowness of work [in securing timber supplies according to prescribed procedures], or with the formalities . . . . [T]hey demand free distribution of forest [materials] at first for soldiers, for the poor and then for everyone. They make it known to any foresters who might be unaware who now has power [*vlast'*], they explain that they—the people—are the power. . . . All matters are decided locally by the people, whether it concerns property,

[37] The terms "obobshchestvlenie" and "sotsializatsiia" are used, usually more or less indiscriminately, in this connection throughout a variety of sources. During early 1918 Faleev bucked this trend by defining the former term in great detail as a kind of compromise measure between local and central control.

valuables or even the life of a person.[38]

If Russia's rural populations were the main focus of attention in connection with forest destruction, they were so because they were both the newest (in the sense of having suddenly accelerated their activities) and the most chaotic threat, not because they were the only one. Authorized cutting by the state continued to take a heavy toll too. And here some new developments were underway. During 1916 the demands of the war economy led to the creation of a number of new economic organs designed to procure and supply timber to high priority users, particularly railroads and factories. These organs included the Special Conference on Fuel (*Osoboe soveshanie po toplivu*), the Central Committee for State Timber Purchases for the Railroads (*Tsentral'nyi komitet po zagotovkam drov zheleznym dorogam*), Local Committees for Regulating Rail Transport (*Poraionnye komitety po regulirovaniiu perevozok po zheleznym dorogam*) and the Field Construction Administration (*Polevoe Stroitel'noe Upravlenie*, or *Lesostroi*). All seem to have had similar purposes and functions and to have raised similar concerns among foresters. The latter, for example—created in February 1916 to supply a variety of government enterprises[39]—was the subject of a report filed shortly before the October Revolution with the central soviet of the Foresters' Union. Over the twenty months or so from its inception to 1 October, 1917, the report claimed, the Field Construction Administration had "cut down no fewer than 50,000 *desiatinas* of forest." Over the next twelve months it was projected that in order for the organ to fulfil its quotas, a further 40,000 would have to go, and beyond that, who knew? "A brief contemplation" of this general process, the report noted, would convince "any genuine forest specialist" of the "vital necessity, of his direct responsibility to come to the rescue of his native forest and to save it from destruction by any means possible." This was especially so for foresters connected with The Field Construction Administration itself, "in front of whose eyes [the country's] forest wealth was dissipating" and who were, allegedly, currently failing more or less completely to take even those minimum measures to ensure forest regrowth following cutting which the Field Construction Administration itself had created and officially adopted. For example, a

---

[38] GARF. Fond 5520, op. 2, del. 16, st. 93.

[39] A. Rerikh, "K voprosu o vozobnovlenii vyrubok proizvedennykh Polevym stroitel'nym upravleniem." GARF 5520, op. 1, del. 34, st.st. 35-36 (dated 1 October, 1917).

regulation requiring a payment of 25 rubles to be made to the local Forest Preservation Committee for each *desiatina* of forest cut down (to pay for the clearing of debris and to otherwise promote natural regeneration) was being completely ignored. Two years after establishment this fund still had not collected even a single ruble. Nor, apparently, had the Field Construction Administration come close to fulfilling its legal obligation to clear up the sites of its own logging operations. In this regard it had not even filed necessary papers. Overall, the report concluded, authorized cutting by state organs, if less widespread and elemental than that being done by peasants all across the country, was little better organized and perhaps just as damaging.[40] Much more evidence in this direction was to emerge over the following months.

The Provisional Government and Forest Conservation

Such then, in broad detail, was the landscape of opportunity and burden presented in 1917 to the would-be creators of a new forestry. For most of that year, the challenge was taken up within the political framework established by the Provisional Government. Thrown together more or less ad hoc, uncertain even of its own legitimacy, and overwhelmed with crowds of vital issues associated with a period of war and political chaos, the Provisional Government would at first glance seem unlikely to have had either the time or the interest to tackle issues of forestry, least of all preservation. On the other hand, it was clear that something had to be done if one of the country's most vital economic resources was not to be compromised. In the event, though short-lived, the Provisional Government oversaw significant developments in forest legislation and organization.

Efforts to preserve or re-establish at least some level of order in the forests came at first in the form of a variety of emergency directives and circulars issued throughout the spring and summer by the Ministry of Agriculture[41]—specifically from two sources therein: the Agricultural Board (*Kollegia zemledeliia*) and the Central Forest Section (*Tsentral'nyi lesnoi otdel*), as the forest department's

---

[40] Ibid., st. 35.

[41] This was under the leadership of A. I. Shingarev, a Kadet, from March to May 1917, and of V. M. Chernov, leader of the SRs from May to August.

leadership was at that time sometimes known.[42] These typically inveighed against illegal and unsustainable cutting and urged conservation while also aiming to establish or reestablish norms for the supply of timber to local populations. By and large, however, and in both connections, they seem to have had little effect.[43]

Much more consequential developments also began that spring. On 21 April the Provisional Government decreed the creation of new local organs called Land Committees (*Zemel'nye komitety*). These were designed to facilitate short term land reform and provide temporary management or resolution of other rural issues until more permanent measures could be organized at the planned Constituent Assembly.[44] Though not intended specifically as forestry organs, the inevitable overlap of land and forest issues assured the Land Committees immediate relevance in this area too. They became greatly more important in forest affairs only a few weeks later following a proposal made by the Ministry of Agriculture to the Main Land Committee (*Glavnyi zemel'nyi komitet*) of a project entitled "Rule on Forest Conservation" (*Pravilo ob okhrane lesov*). Among the Rule's many well-worn provisions intended to stem ongoing forest abuses and promote greater popular interest in conservation, was a more far-reaching and novel proposal to transfer immediately all private and state forests to the control, though *not* to the ownership, of the Land Committees themselves, that is, to local organs of government. This proposal quickly met "serious opposition," however, generally from political *Kadet*s (Russian liberals) and forest industrialists (*lesopromyshlenniki*) who, not surprisingly, "insisted on keeping forest exploitation and [the stockpiling of] supplies in private hands."[45] Among those

---

[42] The term went out of use in the first year or so of the Soviet period but was revived subsequently, probably during 1918; see chapter 5. On the emergency directives themselves see, N. I. Faleev, "Chto sdelano i chto predstoit' sdelat'," 467. See also D. K. Sazhin, "27-fevralia. 27-ianvaria. 27-maia," *Lesa Respubliki* 16 (1918): 840.

[43] See for example,"Polozhenie lesnogo dela v Tambovskoi gubernii s nachala revoliutsii po 15-go maia n. s. 1918 goda." GARF. Fond 5520, op. 2, del. 16, st.st. 90-93.

[44] Land Committees were dominated by Socialist Revolutionaries (SRs), and in some cases by Kadets (liberals). In general, the Bolsheviks trusted only the lower level *volostnye* Land Committees, considering the others hopelessly hostile or "corrupt." The Bolsheviks earmarked them all for abolition in January 1918. On 6 February (old style) they were formally abolished in favor of local Soviets. In fact, Land Committees appear to have persisted for some time thereafter.

[45] N. I. Faleev, "Chto sdelano i chto predstoit' sdelat'," 467.

early lined up against the project in this manner were K. K. Iurenev, then Minister of Transport (who saw a threat to the construction and operation of railroads, which consumed large quantities of timber[46]), State Controller F. F. Kokoshkin, and M. I. Tereshchenko, a sugar magnate who during these months served first as Minister of Finance then of Foreign Affairs. Nonetheless, with only minor amendments, the draft for "Rule on Forest Conservation" was approved toward the end of August and given over to a judicial commission of the Provisional Government for final editing. It was subsequently passed on 10 October under the altered title, "Temporary Rules on the Conservation and Cutting of Forests" (*Vremennye pravila ob okhrane lesov i ikh rubke*).

The success of the decree and the general shift toward mostly local forest administration appear to have had two main causes. First, at the time, local control was the only realistic solution. The Provisional Government's chief priorities lay elsewhere. In particular there was a desire to continue hostilities against Germany and her allies in fulfilment of the nation's war treaties and other foreign commitments. This meant that where possible less pressing matters would have to be shelved until the end of the war. And since local control of the forests was in any case already more or less a fact—effected by the ongoing peasant seizures, the beating and expulsion of forest personnel, and so on—the decision appears to have been made in essence legally to sanction (and thus introduce at the least the appearance of legal order to) the existing state of affairs. At the same time, even the *central* offices of power over the rural economy had come under the control of new blood personally rather favorable to peasant politics. In particular, V. M. Chernov, a founder and leading figure of the pro-peasant Socialist Revolutionary Party, headed the Ministry of Agriculture from May to August. "Only when . . . Chernov received the ministry's portfolio" it was noted the following year, was it "possible to speak of a beginning of forest refom."[47]

---

[46] Ibid.

[47] Ne-Lesovodov, "Lesnaia revoliutsiia," 850-51. Chernov's appointee P. A. Vikhliaev, it seems, did most of the actual relevant work, however, since Chernov himself was focused on other matters most of the time. Vikhliaev is described by Ne-Lesovodov as an SR both of "extreme right" and "extremely variable" persuasion, and also as a former Kadet who still often evinced "kadetishness" in some of his work. Vikhliaev's attitude to forest reform, at least according to this source was "hard to pin down." Vikhliaev, it is claimed, "failed to understand" forest affairs in general and "embarassed himself" in this regard when speaking at forest meetings during the Provisional Government era. Most of this is likely pro-communist

Chernov, moreover, was only one of many SRs to achieve high office during this period, compounding the drift toward peasant politics. On the other hand, this tendency was countered to some degree by an influx also of individuals more willing to promote the priorities of the center rather than of the localities, especially within the Forest Department itself. Among these was N. A. Kuznetsov. A strong advocate of the "organic centralism" discussed in chapter 2, [48] Kuznetsov joined the department in April 1917 along with several other new personalities. He later described this influx and related changes as having "freshened [the Department's] atmosphere, breathed into it a living spirit, a restless spirit.[49] Pro- centralizers such as Kuznetsov were for the moment, however, outnumbered and outranked by SR and other pro-localism persons. It seems, further, that many individuals were affected by the general political environment—waxing localist under the Provisional Government and then centralist under the Bolsheviks. N. I. Faleev, who was beginning at this time to emerge as the major forestry figure overall (at that time within the Agricultural Board), was himself far from the centralizer he was to become shortly after October 1917, as will be seen elsewhere. Overall, then, while both new legislation and events in the countryside had significantly decentralized forest control, Russia's leading forestry personnel remained split about whether or to what degree this was a desirable development.

The "Temporary Rules on the Cutting and Conservation of Forests"[50]

Framed in the expectation that they would soon be superseded by the creation of "new special organs of forest conservation" (article 1), the temporary laws nonetheless introduced some important changes and can usefully be seen as

---

bluster more than real information. Vikhliaev apparently wrote an unpublished work "Socialization of the forests," about which no information is available.

[48] For example, after October 1917 Kuznetsov bitterly criticized early drafts of Soviet forest law drawn up by N. I. Faleev on the grounds that these did not provide sufficient power to the center. See chapter 5.

[49] GARF. Fond 5520, op. 2, del. 1, st. 8.

[50] *Sobranie uzakonnenii i rasporiazhenii pravitel'stva* (Moscow, 1917), st. 1990.

a significant first step on the road toward a state-owned and managed forest economy. As noted, the new laws granted important new powers to local organs—specifically to the Provincial (*gubernskye*) and District (*uezdnye*) Land Committees—to regulate and oversee the cutting of timber and ensure the "preservation and proper use . . . of all types of forests, private and public [*obshchestvennye* (article 2)]." To this end the decree provided for the creation of dedicated forest personnel within each Land Committee (articles 1-2). In addition to the duties and powers already given at their establishment on 21 April, both Provincial and *uezdnye* Committees were empowered to "make declarations to the Minister of Agriculture regarding the necessity of declaring forest areas "protected" or "water-preserving" on the basis of article 741 of the existing Forest Charter and to carry out checks to see if forest owners were in compliance with the Statute of 1888 upon which this part of the Charter was still based. They were also charged with granting certification (*udostoverenie*) to cut.[51]

Marking something of a break from earlier legislation, the new law also required that all cutting permissions granted to owners or managers of forest areas by Forest Preservation Committees (under the conditions of an 1892 Amendment to the 1888 Statute) were to be presented to *uezdnye* Land Committees for reevaluation (article 6). In the event that those presenting were not satisfied with the decisions made for their lands at the local level, appeals could be made to the Main Land Committee (*Glavnyi zemel'nyi komitet*), and beyond that to the administrative courts (article 11). Provincial Land Committees were also empowered to grant or deny permission to clear forest for agricultural use (article 13), and were urged to "use all legal methods" to eradicate unauthorized or otherwise destructive use of the forests.[52]

On the other hand, and of some significance, while article 5 of the "Temporary Rules" required that private or public (*obshchestvennye*) forests could be cut only upon receipt of authorization by *uezdnye* Land Committees, it was stated that this was not so regarding some of the procurement organs noted earlier, including the Special Conference on Fuel, Central Committee for State Timber Purchases for the Railroads, or Local Committees for Regulating Rail Transport. These needed only to inform their *uezdnyi* Land Committee "about the

---

[51] It is unclear from the printed law whether these certificates were mandatory or optional.

[52] See also *Lesa respubliki* 1 (1918): 43-44.

cutting they intended to do." They would then "quickly receive authorization."[53] Cutting permissions granted to these organs could not be revoked by local organs. Even in the case of significant damage to the forest, or to the future timber supplies of these or other important industries, local organs were granted only the power to bring matters to the attention of the Minister of Agriculture (article 8) Though important limitations had existed also under the tsars on the authority of the forest planning apparatus to protect forests in the face of economic imperatives, the relatively independent status of economic organs (the above-mentioned Field Construction Administration included) was a new and growing trend that climaxed after the Bolshevik Revolution with the effective split of forest affairs into two separate branches, each with its own organs and jurisdiction. One branch—planning and administration (including conservation)—would be known as "forest economics" (*lesokhoziaistvo*) and was the domain of the agricultural organ; the other—known as "forest industry" (*lesopromyshlennost'*) and entailing physical procurement, distribution, processing, and so on—was to be overseen by the chief economic organ, the VSNKh. This is taken up more thoroughly in chapters 6-7.

"Instructions" for Implementing the "Temporary Rules"

On 1 November (Old Style), a week after the Bolshevik seizure of power, the agricultural organ, still operating on the momentum established during the period of the Provisional Government (and also within its framework) issued a set of instructions governing the implementation of the "Temporary Rules" (*Nastavlenie po primeneniiu "Vremennykh pravil ob okhrane lesov i ikh rubke*").[54] Above all, these detailed the division of relevant tasks and authorities among the four levels of Land Committee: *volost'nye*, *uezdnye*, *gubernskye*, and *glavnyi* (main or chief).

The opening section, focused specifically on forest protection, offered some seemingly useful provisions. It established that the "first act" of the Provincial and *uezdnye* Land Committees should be to organize a Forest Supervisory body [*Lesnoi nadzor*] to "provide the population with all types of assistance in the area of forest economy and forest protection." Specifically, this

---

[53] *Sobranie uzakonnenii i rasporiazhenii pravitel'stva* (Moscow, 1917), st. 1990, articles 4-5.

[54] I have here relied on a printed copy of the *Nastavlenie* deposited in the Foresters' Union's archival holdings for 1917. GARF. Fond 5520, del. 34, st.st. 55-65.

would involve overseeing and reporting back to the relevant Land Committees on all felling, stockpiling, and exporting practices in private and public forests, in accordance with the forest regime already outlined above. Members of the Supervisory were required to be resident in the area where they worked and in possession of "forest-technical knowledge" [*lesotekhnicheskye poznanye*]. Where evidence was discovered of "devastating cutting" [*opustoshitel'naia rubka*] or any other unauthorized or illegal forest clearing, they were to carry out all necessary "technical research" and, if required, also to order temporary halts on all cutting. The Supervisory was also empowered to initiate proceedings against forest law violators and to bring evidence against them in court (articles 1-6). Other clauses made similar provisions regarding the grazing of animals in the forests (generally illegal in areas of young growth or where soil erosion was a concern) and other secondary forest usage, and generally outlined procedures for assessing or appealing grants or denials of cutting, grazing and other forest use permits.

The "Instructions"—here essentially reiterating the existing tsarist Forest Code—also outlined which types of forest areas should be considered for designation as either "protected" or "water preserving" (*vodookhrannye*). The former included those "holding together loose soils" or percolating sands (*sderzhivaiushchye sypuchye peski*) in general or in a variety of specified instances where erosion deposits might negatively impact navigation, water supply, railroads, agriculture, and so on (article 85). The latter included "forests and shrubs, the destruction of which threatens to lower the water level in springs and headwaters of rivers and streams" (article 86). Following the designation of a forest area as either "protected" or "water preserving," its conversion to any other use was "absolutely forbidden" without exception (article 87), essentially a restatement of a corresponding clause of the 1888 "Statute." While the Forest Supervisory or any level of Land Committee was required to report on the existence of areas matching these descriptions, and to carry out relevant research therein, the decision actually to declare them protected rested exclusively with the Minister of Agriculture (article 61). The Forest Supervisory did, though, have the authority temporarily to ban cutting in such areas pending the Minister's decision.

These decrees were not, however, the strong or effective statement of forest protection they might at first glance seem. Moreover, many of their faults were the same ones identified by Zaitsev and others prior to the First World War (see chapter 1). "Clear-cutting" (literally, "devastating cutting," *opustoshitel'naia rubka*), though declared illegal in general, still received no clear definition.

Remaining in effect also was the rule that any cutting carried out "entirely in accordance with an economic plan confirmed in a lawful manner" could not be deemed clear-cutting. It has already been mentioned, further, that the laws held little sway over the activities of state economic organs, which needed only to inform the Land Committee "about the cutting they intended to do" before "quickly receiv[ing] authorization" which could not be revoked except by central dictate.[55] There were other loopholes in the law too. *Uezdnye* Land Committees, for example, were obliged to notify the Forest Supervisory of proposed cutting on public or private land only when "large" areas were involved (which term was not defined, articles 64-65). Exemptions were also allowed from certain regulations, such as the ban on grazing animals in sensitive areas, if enforcement were to be judged likely to "bring ruin" on the landowner or local economy (article 44). Ultimately, the new laws seem to have set as a priority the organization of felling under a stricter—yet also more decentralized—chain of command, rather than the absolute reduction of cutting, *per se*. In any case, the virtual absence of central power and legal order at this time rendered these laws somewhat academic for the time being.

Other Developments

Aside from these legislative efforts, and in search of more "socialistic" and "rational" forms of forest exploitation, experiments were made at the same time to create and deploy collectives of *lesnye artely*—small-scale quasi-socialist associations of rural workers through whom some members of the forest apparatus hoped conservation, regeneration, and other vital forest work could be carried out. These efforts also seem to have enjoyed little success.[56] Almost immediately after the Bolshevik Revolution they fell victim to criticisms they

---

[55] *Sobranie uzakonnenii i rasporiazhenii pravitel'stva* (Moscow, 1917), st. 1990, articles 4-5. Even in the case of significant damage to the forest, or to the future timber supplies of these or other important industries, local organs were granted only the power to petition the Ministry of Agriculture, not to halt cutting (article 8).

[56] The *lesnye artely* enjoyed only a short existence—or at least a short period of favor by the new forest administration—peaking during the winter of 1917-1918. They were already clearly out of favor, and largely out of action, by the passage of "On Forests" in May 1918. *Lesa respubliki* frequently refers to them during 1918 as "capitalistic," or infiltrated and used by "capitalists" and "plunderers."

were "vehicles" through which forest privateers and capitalists continued their influence and operations. They were, accordingly, quickly de-emphasized in favor of directly drafting peasants into forest work via local Land Committees or, where they had been established, local Soviets.[57]

During the second half of 1917, here and there *sovety lesnichestv* (literally "Soviets of the managed forest unit") also began to emerge. Comprising local residents, *volostnye komitety* and other small "land organs," these seem typically to have been poorly organized and had little real influence on forest matters.[58] They are perhaps most interesting for the hostility they received from the Foresters' Union, which adamantly opposed efforts to grant what it saw as groupings of unqualified amateurs any increased prestige or control over the forests. For similar reasons, the Union also criticized, though perhaps less aggressively, the transfer of control to Land Committees granted by the legislation noted above.

Finally, even before the publication of the "Temporary Rules on the Protection of Forests and their Cutting" efforts were already underway within the Forest Department to create a permanent and more comprehensive code of forest legislation. According to Faleev, this effort was already being directly along clearly socialistic lines, with forest specialists and "representatives from lumber factories [*drevoobrabativaiushchye zavody*], banks [and] chambers of commerce [*torgovye palaty*]" daily being invited to discussions at the Department.[59] This, in fact, was the early history of subsequent Bolshevik legislation.

* * *

Given the transitory nature not only of the Provisional Government itself, but of the entire historical period it represents in Russia's development; and given also the tenacity and transforming nature of the Bolshevik regime that followed, one might be tempted to write off as historical cul-de-sacs also the

---

[57] See for example, "Lesnaia sektsiia: itogi rabot lesnoi sektsii" *Lesa respubliki* 17-19 (1918-1919): 901-04. On efforts made following the Bolshevik Revolution to replace *lesnye artely* with *lesnye kommuny* see *Lesa Respubliki* 8-9 (1918): 476.

[58] N. I. Faleev, "Chto sdelano i chto predstoit' sdelat'," 468-69.

[59] Ibid., 469.

events and achievements of this period in forestry. In fact, nothing could be further from the truth. The Bolsheviks, to be sure, were to bring major changes to forest management and protection as to other areas of life in the lands of the former Russian Empire. But the period between the February and October Revolutions laid important foundations for much of what was to come. In particular, the collapse of political power associated with the first Revolution cleared away the old social, legal, and political edifice, making substantial change possible. Simultaneously, however, it also unleashed the furies of the peasantry against forests and foresters alike, setting in motion a ruinous hurricane of centrifugal forces the Bolsheviks were to inherit and would long struggle to contain and reverse. The legislative efforts undertaken during the era of the Provisional government, though less fundamentally important, matter also—primarily in that they effected a legal transfer of authority over the forests to local organs. Thus, by the time the Bolsheviks took center stage both "facts on the ground" and legislative developments had combined to create a decentralized forest *fait accompli*. Neither the real nature, nor the full ramifications of this, of course were necessarily obvious at the time—either to foresters convinced of the possibilities of local-central cooperation, or to the Bolsheviks for whom the compatibility of central and local interests under conditions of socialist revolution should theoretically have merged seamlessly.[60] The real struggle, then, was only beginning.

[60] Because they transferred control of forests to the population via its local land organs, the "Temporary Rules" and "Instruction" of October-November were summarized in late 1918 as having represented an important, if modest "first step on the road to socialization." See D. K. Sazhin, "27-fevralia. 27-ianvaria. 27-maia," 842.

IV

The Bolshevik Revolution and Forests:

From October 1917 to the Issuance of the "Basic Law on Forests," 30 May, 1918

A momentous event in virtually every aspect of Russian history, in forestry the Bolshevik Revolution was in some ways more facilitative than creative. Of the four major forest-related initiatives to be approached following October 1917—total nationalization, organizational centralization, the introduction of scientific planning on a national scale, and reorganization of the relationship between specialists and politicians—all but the last had been among the key reforms called for by a variety of pre-Revolutionary foresters. N. I. Faleev, by now busy writing "On Forests," made a similar point himself the following year, noting that the Revolution had simply "allowed for the realization of all the wishes of the leaders of forest politics which had previously been set out and anticipated."[1]

If the Revolution did not create *de novo* its own set of concepts about forest organization and use, it nonetheless had a critically important effect in shaping the direction of existing trends. Above all, it assured that momentum toward state ownership and planning would be steered as much as possible in the direction of central rather than local control, and thus toward a single plan rather than many. The one truly Bolshevik innovation, as indicated, involved the relationship between specialists and politicians. Committed as they were to scientific planning and resource use, the Bolsheviks clearly were to need large numbers of technical experts. As will be discussed more fully below, however, an urgent priority was quickly established to place these under strict political control, explicitly demoting specialists as a group from the relative autonomy and considerable prestige they had previously enjoyed to the status of "organ" of the state. Thus, though foresters' plans came increasingly to be embraced by the new political authorities, foresters themselves were not destined to control efforts to implement them.

All four programs here mentioned—nationalization, centralization,

[1] N. I. Faleev, "Chto sdelano i chto predstoit sdelat'," 471.

planning, and the imposition of political control over technical experts—overlapped considerably. They were soon to be enshrined in the "Basic Law On Forests," the Bolsheviks' first major forest code, issued in May 1918. "On Forests" itself will be examined in the following chapter. Here, an overview will be given of the seven months or so separating the Revolution from this decree. During that time the Bolsheviks made their early efforts to plant within their chaotic inheritance the seeds of the new forestry already outlined. Over the course of these few months, also, can be glimpsed the entire range of forest-related problems the Bolsheviks were to struggle against for many years. While some of these problems, such as those caused by the ravages of the First World War and the Civil War, can rightly be evaluated as temporary and unavoidable, others, it will be seen, were more specific to the nature either of the early Soviet system itself or, more importantly, to the basic structure of the Russian society of which the Bolsheviks had assumed leadership.

Nationalization and Centralization

The Bolsheviks wasted no time at all in ending private property in land. The "Decree On Land"—issued on 26 October, 1917 at 2 AM, literally hours after the seizure of power—abolished without compensation all private holdings and turned land in general over to the disposal of local Land Committees and Soviets of Peasants' Deputies (at the *volost'* and *uezd* levels, respectively).[2] A more thoroughgoing socialization law—of all land, minerals, waters, forests, and "living forces" (*zhivye sili*)—was issued the following 27 January under the title, "Basic Law on the Socialization of Land." Not the implementation of rational use policies but winning over the peasantry to the cause of Bolshevism was the immediate concern of these decrees. Lenin, after all, had won considerable support, and hoped to win much more in the coming critical struggles against political enemies, by making promises to the population of "land, peace, and bread." Thus the laws focused primarily on peasant farms, not forests. Article 5 of "Basic Law on the Socialization of Land" in fact explicitly deferred a proper treatment of forests (along with waters, minerals, and "living forces") to separate "basic laws" subsequently to be issued. Similarly, both these land laws promoted

---

[2] The Bolsheviks distrusted higher level Land Committees, considering them hopelessly tainted with "bourgeois" tendencies.

what was by now increasingly being called "socialization" rather than "nationalization" of land.[3] This meant that they made considerable concessions to localist and "petty bourgeois" sentiment at the expense of central control (see chapter 5).[4] Full land "nationalization"—emphasizing the nation over local communities—was forthcoming only on 14 February, 1919, some sixteen months after the Revolution, and nearly ten months after the issuance of the "Basic Law On Forests" had already established clearly centralized and national principles specifically for forest management (also see chapter 5).

As noted, the October Revolution brought to power individuals committed generally to centralization. (We might say, alternatively, that it brought to power persons convinced of the correctness of only one basic set of policies and practices, and of the necessity and justification of bringing these to bear as widely as possible. In either case both the means and result involved centralization.) Lenin himself had made this priority clear in a circular of 4 April, 1918, stating unequivocally that "all forests constitute the property not of any village, province, or region, [but] represent a general-public [*obshchenarodnyi*] fund and cannot in any case be subject to any division."[5] In this environment heterodox expressions of localism and geographical or political fragmentation, including in relation to forests, were bound to be viewed dimly. The spirit of the day was adroitly reflected in the comments of the pro-centralizing forester, T. N. Ponomarev: "[T]he local population's [desire] . . . to take forest affairs into its own charge and . . . for its own interests hinders [the organization of] forestry on a statewide

[3] Though these terms had been used more or less interchangeably in many cases prior to October 1917, thereafter they began to assume more clearly distinguished meanings. "Nationalization" emphasized central control and national planning, whereas "socialization"—of which these land laws were an example—offered greater freedoms to local organs and patronized localist thinking.

[4] Indeed, the first decree in particular was widely interpreted in the localities more in anti-governmental terms than as part of a change of government or land law. In forest matters, large numbers of peasants reputedly saw—or chose to interpret—the decree as abolishing forest management by state bureaucracies altogether (primarily the hated *Upravlenie zemledeliia i gosudarstvennykh imushchestv* which had been largely unaffected by the February Revolution. See "Tsentral'noe Upravlenie Lesov respubliki," *Lesa respubliki* 5-7 (1918): 341; and ibid., 1 (1918): 43-45.

[5] *Lesnoe khoziaistvo SSSR za 50 let, 1917*-1967, 16-17.

scale." It would thus have to be resisted.[6] Another forester P. Serebrennikov—writing in the January 1918 issue of *Lesnoi zhurnal* made a similar point:

> State forest organization, in the broad sense of the word, must encompass all the territory of the country, establish an economic plan entirely satisfying the demands of the whole country, of all the people. Such a coordination of the economic plans of each separate forest area [*dacha*] is at the present moment for the [new] government one of the most important tasks.[7]

"All the artificial juridical borders among forest areas," he predicted, "would fall."[8] "State forest organization" would from now on have to "keep in mind the interests of the wide democratic masses of the population, not only the provincial [*mestnye*] populations, but rather the entire masses living on the whole territory of the country."[9] The "supreme oversight of the state forest system," he concluded, "must be concentrated in the central administration, for the purpose of unifying the entire forest politics of the country."[10] Given current forest realities, these were revolutionary ideas indeed.

The perceived need for the center to wrest control of the forests from local populations followed not only from the irresponsible actions of the latter at the present moment but represented—both for the new regime and the apparent majority of foresters overall—a general principle without which neither rational use nor adequate forest protection were possible. Only centrally-coordinated forestry, for example, could ensure adequate forest materials for the benefit of the whole nation, rather than for the few, for generations to come. In certain areas (the far north and Siberia) forest resources were superabundant, requiring increased exploitation; in others there were growing local shortages and a critical need for strict preservation and regeneration. Thus, national, centrally-coordinated, and long-term perspectives were indispensable. Rational forest management also required considerable financial outlays—for replanting, forest care, the construction of roads and other transportation infrastructure. And there was a need for technical expertise, not the spontaneous action of untrained

---

[6] GARF. Fond 5520, op. 2, del. 18, st.st. 95-96

[7] P. Serebrennikov, "Zadachi gosudarstvennogo lesoustroistva," *Lesnoi zhurnal* 1-2 (1918): 8.

[8] Ibid.

[9] Ibid., 13.

[10] Ibid.

persons. On all these counts, the unfolding whirlwind of centrifugal forces and localist particularism then sweeping Russia could offer little to those seeking to create Bolshevik forestry and would clearly have to be controlled and reversed.

Effecting such changes on the ground would not be easy, however. A first priority was to calm somehow the waves of forest destruction still unabated during the winter of 1917-1918. Unfortunately, Russia was at that time lacking almost every semblance of political organization, authority, and stability; it was also coping with the aftermath of the First World War and making preparations to fight what was inevitably going to be a bitter and difficult struggle against a variety of counter-Revolutionary forces. Not surprisingly, the country had few resources at its disposal to police the forests or implement planned use. Consequently, much reliance was made at first on simple exhortation, usually in the already-familiar form of circulars, several of which were issued by the new forest authorities and the *Kollegia Zemledeliia* during this period. Predictably, these had no noticeable effect.

Despite numerous distractions, more systematic efforts were also undertaken early in 1918 aimed at creating a functioning organizational link between local and central forest affairs. An early step in this direction was the new forest department's publication on 12 January, 1918 of the "Temporary Decree on Forest Administration" (*Vremennoe polozhenie ob upravlenii lesami*). This provided for the establishment of Forest Sections (*Lesnye otdely*).[11] The Sections were supposed to function as representatives of the emerging central forest organ deployed in local organs for the purpose of helping unite the former and latter. This they would do, it was hoped, both by relaying information on local conditions to central planners and by promoting central imperatives in local environments. At their apex, a Central Forest Section [*Tsentral'nyi lesnoi otdel*] was established.[12] (This took over the functions of old Forest Department and came under the ultimate jurisdiction of *Narkomzem*, the chief agricultural organ. Further reorganization of forest administration was soon to come, however. The term "Forest Department"with or without capitals—continued to be widely used as a generic appellation for the evolving central forest administration and will thus be preserved throughout this book.)

At first, Forest Sections functioned primarily within existing Land

[11] These had first been outlined in a circular of 15 December, 1917.

[12] N. I. Faleev, "Chto sdelano i chto predstoit sdelat'," 472.

Committees; but by early 1918 they were already coming increasingly to rest within the growing number of provincial Land Soviets (*Zemel'nye sovety*). Land Committees, moreover, had by January already been targeted by the new regime for abolition—a development which was made formal later that month by the "Basic Law on Land."[13] In this connection, at the end of January, shortly after the Third Congress of Soviets, the Main Land Committee was succeeded by the Main Land Soviet (*Glavnyi Zemel'nyi Sovet*).[14] At about the same time, the Agricultural Administration (*Upravlenie Zemledeliia*) was also abolished by a temporary "Decree on Local Forest Administration" (*Polozhenie o mestnom lesnom upravlenii*). Its responsibilities were transferred at first to the Forest Sections, and shortly thereafter to Soviets of People's Deputies (*Sovety Deputatov*).[15] After 23 February, 1918, the Forest Sections inherited also the functions of the old Forest Preservation Committees (*Lesookhranitel'nye komitety*), abolished that day by decree of the *Kollegia Zemledeliia* (Agricultural College[16]). Insofar as each of these developments furthered the growing trend away from pre-Revolutionary forms of local administration toward Soviet control they can be seen as steps in the direction of overall political centralization under the new government. In reality, however, this was not always so, since the relationship between the Bolshevik authorities and provincial Soviets was itself often highly tenuous. Not Bolsheviks but SRs dominated many local Soviets, especially in rural areas where forest politics were most relevant. The achievement of a coherent chain of command in forest (as in other) affairs still lay far ahead.

During April 1918 further clarification and development was given to the emerging organizational structure of forest management and in particular to the key role therein of the Forest Sections. A session of the *Kollegia Zemledeliia* held that month stated its primary intention to combat the "bureaucratic" and "excessively centralized" nature of previous forest administration and to organize the new one in more "expedient" and "systematic" fashion. In line with what has been said already about the envisioned "organic" dynamic between central and local authority, the session outlined a forest administration whose general structure, priorities, and overall plan would be established by the center (in the

---

[13] In fact, numbers of Land Committees continued in existence for some further time.

[14] "Bor'ba protiv Osnovnogo Zakona," *Lesa respubliki* 8-9 (1918): 461.

[15] See, for example, D. K. Sazhin, "27-fevralia. 27-ianvaria. 27-maia," 843.

[16] Located within the main agricultural organ, *Narkomzem* (People's Commissariat of Agriculture).

form of a revamped forest department) while leaving considerable initiative and flexibility in the hands of local authorities (especially the Forest Sections). This flexibility was to be confined within clear parameters, of course: local initiative was to be of a technical rather than executive or political nature.[17] And ultimate control over the education of forest personnel, their appointment, transfer, and dismissal were to remain under central control, at least in law. (It was in this context, in fact, that the subordination of expertise to politics was largely conceived.) More broadly, to the central forest administration now taking shape (the old Forest Department) was entrusted what has already been noted as "forest economics" (*lesokhoziaistvo*)—"the general management of forest affairs, the working up of legislation and general measures, the putting together of a 'general state' plan for the forest economy, working out state estimates of demand for forest [materials] and the satisfaction of these, [and] general oversight and control."[18] All these ideas were given legal form in April in a "Regulation Regarding the Authority of Local Forest Sections" (*Polozhenie o kompetentsii mestnykh lesnykh otdelov*.)[19] Further information on the function of the Forest Sections was forthcoming the same month. Sections were to inventory forests under their charge, assess local state, industry, and popular timber needs, and help physically supply those needs, thus establishing the bases of a planned forest economy. They were also to be responsible for day-to-day forest conservation and regeneration work, including planting, clearing, fencing, and so on. Housing, equipping, and supplying forest guards was also made their business, as was "resolution" of routine complaints and appeals lodged by forest guards. (Given that Forest Sections' personnel seem in at least some cases to have comprised local peasants and SRs hostile to the old forest guard, this latter fact may well have had the awkward consequence that guards expelled from the forest were in some cases required to take their grievances to bodies composed of their tormenters!) Forest Sections were also charged with uncovering forest law violations, initiating prosecution, and with issuing and enforcing local fire-

[17] "Kompententsiia lesnykh otdelov i lesnogo departamenta," *Lesa respubliki* 2 (1918): 119-20. See also N. I. Faleev, "Chto sdelano i chto predstoit sdelat'," 472.

[18] N. I. Faleev, "Chto sdelano i chto predstoit sdelat'," 472.

[19] Ibid.

prevention measures.[20]

These considerable responsibilities were spelled out more precisely at about the same time in a list of forty-five explicit duties, some of which appear at first glance to allow these organs significant autonomy from the center. These included the right to establish the methods for carrying out cutting and natural and artificial [forest] regeneration. The "scale and character" of different types of cutting (*prorezivaniia* and *prochistki*), as well as norms for "secondary" forest use (presumably including the gathering of dead wood, berries, mushrooms), could also be set by Forest Sections.[21] And in cases of "extreme necessity" Forest Sections were even entitled to "make alterations in the methods, time and place" of planned or ongoing cutting." In fact, neither these nor any other grants of authority to the Forest Sections or other local bodies in any way altered the overriding commitment to the center as the proper and only source of basic priorities and methods. The above-noted "deviations from an established [forest] plan," for example, were permissible only if they "did not involve any raising of the level of [forest] use" and did not "threaten damage to newly-planted areas." They must also be accompanied by a "complete guarantee" that the cut area be regenerated with the same major tree species that had been felled.[22] And in all other ways they could not violate the basic priorities and principles of a centrally-planned forest economy at that time still in the process of being properly worked out.

The commitment to the center holds true even regarding several explicit statements by the central forest apparatus regarding measures and intentions promoted as "decentralization." In order for this assertion to make sense, it must be borne in mind that by early 1918 "centralism" was more and more becoming simply a synonym for the perceived imposition by the *tsarist* Forest Department of unfair authority along wrong-headed lines. Basic to the thought processes of the emerging new regime was the notion that the vast majority of persons (the working masses) shared fundamental common interests, in forestry and in general, that these interests were best expressed through socialist organization, and that sooner or later a large enough majority would realize these truths. Since they were truths—since there was essentially a single correct way to organize forest policies—they should and would be adopted by everyone. They must also

---

[20] *Lesa respubliki* 1 (1918): 35.

[21] "Kompententsiia lesnykh otdelov i lesnogo departamenta," 122-26.

[22] Ibid.

be properly organized and coordinated. This meant centralization, even if construed otherwise. This was, again, simply the same "organic centralism" deriving from pre-Revolutionary forest debate. In retrospect, of course, it is easy to see that such a common vision was entirely lacking, and that state and population hoped to pursue very different trajectories. At the time however, this was likely not so obvious.

Beyond Forest Sections, other local organs were envisioned too, to be organized and run in cooperation with the Forest Sections, and apparently overlapping with them also. "Special technical organs" were planned, for example, to be charged with carrying out localized "technical tasks," a major component of which was to be forest preservation. By this, of course, was meant the actual, physical work of preservation; no real authority was imagined for these organs. Indeed, they are to be understood within the context of the demotion of experts and specialists already noted.[23] Technical organs were to comprise a *lesnoi instruktor* (forest instructor) and *tovarishch lesnichii* (comrade forester). The latter, yesterday's *lesnichii* but with suitably revolutionary moniker attached, remained the highest ranking and best-educated forest specialist, reporting directly to the central forest administration. Unlike the "learned" (*uchenyi*) forester of the pre-Revolutionary period, and again in search of better central-local coordination and cooperation, the *tovarishch lesnichii* would be required to live in the forest area to which he was assigned and establish close relations with the local population.[24] He was to be supplied with "at least five" assistants aged between 16 and 20.[25] Directly under the *lesnichii*, the *lesnoi instruktor* was to be chosen by the local Soviet and required to pass a special examination under the auspices of the local Forest Soviet (see chapter 3). The successful candidate was to monitor and gather information on conditions in his assigned forest, including violations of forest law. Aside from attaching the prefix "comrade," the Bolsheviks invented several other more revolutionary- and egalitarian-sounding names for forest personnel. The word "chiny" (rank, or ranked official)—reeking of tradition and hierarchy—was dropped outright. All forest personnel, regardless of background or qualifications, were now to be addressed in general as

[23] Ibid.

[24] *Lesa respubliki* 1 (1918): 36-37.

[25] Ibid., 39.

"government forest specialists."[26] Several other similar name changes were also effected.[27]

Creating new organs, positions, and names was, of course, a relatively easy game. Peopling these with persons loyal to the center was far harder. If by the end of the 1920s—under Stalin—the Soviet state was both able and willing to impose its will upon the population (via the collectivization of agriculture in particular), this was not so in 1917 or 1918. As with efforts to reduce unauthorized cutting, in many cases simple persuasion, such as by circulars, had to suffice. One issued 7 December, 1917 (over the signature of People's Commissar A. L. Kolegaev on behalf of the "Forest Department"), for example, stressed the confusion that occurred when "other organizations" than Land Committees (within which Forest Sections were then being deployed) claimed control of the *lesnichestva*. The circular insisted that no such competing claims could be countenanced. Similarly, even the Commissars appointed by local soviets to oversee forest administration needed constant reminding that they had no authority to command or dismiss *lesnichii*, but were required simply to report any problems with personnel either to the local Land Committee, Soviet, or to the People's Commissar for Agriculture (*Narodnyi Kommissar Zemledeliia*), on the understanding that these latter would themselves take appropriate action.[28] (This was clearly also in response to growing concerns, already noted, that local organs and populations often failed miserably to provide any level of forest protection at all.)

Response to another circular—this one issued 15 December, 1917 by A. L. Kolegaev and N. I. Faleev on behalf of the *Kollegia Zemledeliia*—exampled the problems involved in promoting forest centralization, or even a coherent chain of command. Addressed "to all administrative managers [*nachal'niki upravlenii*], to Land Committees, and to *uezd* districts [*uezdnye okruga*]," the circular noted the imminent "reorganization of the forest administration and assumption by the government of control over [previously] private forests." It called in this light for "immediate negotiations with the corresponding provincial Land Committees regarding the formation within each latter of a [F]orest [S]ection" (as already outlined). To these latter, the circular stated, were to be transferred "control . . . of

[26] N. I. Faleev, "Izmenenie polozheniia pravitel'stvennykh lesnykh spetsialistov," *Lesa respubliki* 3 (1918): 165-66.

[27] For further examples, see ibid.

[28] *Lesa respubliki* 1 (1918): 44-45.

all forest units, private and state, along with their staffs of state officials who will remain subordinate to the central forest organ . . . ." Allocation of forests and materials, and the general order of forest affairs were in all other respects to remain in accordance with the "Instruction" of 1 November (see chapters 2-3).[29] The results of such negotiations, reported the following month, were as follows: only fourteen of twenty-two respondent administration indicated a general willingness either to carry out or even negotiate such transfers. Eight others rejected the whole idea outright, ostensibly on the grounds that it would bring into positions of influence inexperienced or unreliable personnel.[30] Similarly, an unspecified Land Section (*Zemel'nyi otdel,* the agricultural equivalent of a Forest Section, also situated within local Land Committees) in Olonets province, when ordered to cooperate with such transfers in its area of jurisdiction, not only explicitly refused, but ruled instead to increase the amount of forest under its own jurisdiction and then to "transfer of all of this to [its own] exclusive control."[31] In both these cases, and in numerous others, beyond harsh words there was little that could be done.

There were of course, varying degrees to which centralization could be established or practiced. During the first few months after the Revolution several important figures, while totally opposed to the localist impulses of many Social Revolutionaries and the peasant population in general, mindful perhaps of earlier dissatisfactions with the "aloofness" of the old Forest Department, were clearly reluctant to go too far in the opposite direction. The same Georgii Vysotskii who back in 1905 had called for a locally-controlled forest nationalization (see chapter 2), though now apparently more favorable to central control, still professed the "reasonableness" of localized forest administration "in exceptional cases," primarily "in the south, on the steppe, where, due to the unsuitability of soils . . .

---

[29] Ibid., 29.

[30] Ibid., 29-32.

[31] The actions and proposals of this Forest Section were condemned also in October 1917 at a congress of Land Sections held in Petrograd. But beyond censure, there was little that could actually be done at this point (or later). See Al'fa, "Ne priznaiushchye kommunisma," *Lesa respubliki* 16 (1918): 863-65.

forested areas are few and far between and in many cases are also very small."[32] That this was still a more clearly centralist agenda than Vysotskii had supported back in 1905, however, is made clear by his assertion that nationalization "must be carried out in a *centripetal direction* (toward the center) with the *participation* therein of representatives of the interests of the local populations run according to special laws [emphases in original]."[33]

As well as numerous disappointments and challenges to Bolshevik efforts to gain influence over rural bodies, there were also signs of progress. For one thing, the figures noted above regarding the willingness or otherwise of local Land Committees to hand over authority to Forest Sections can, of course, be read optimistically (nearly two-thirds indicated a willingness to cooperate), or pessimistically (more than one-third refused to cooperate). This seems, in fact, to have been about the proportion of provincial cooperation with the Bolsheviks overall during the first year or so after the Revolution, provided it be borne in mind that "cooperation"—like the lack thereof—was never either constant or total in any arena. Here and there reports were circulated also of specific "successes" in establishing local Soviets friendly to the Bolshevik center. The Foresters' Union's Regional Congress held in Moscow in January, for example, heard a report from Kherson province that "the forest apparatus had been organized in accordance with the new Soviet authority already during January."[34] In Tver province it was also reported, the new authorities had already "liquidated" the Land Committees. And there were other similar reports.

Who Will Lead? Specialists and Politicians

In the forests and across the countryside, the ongoing process of Sovietization—in this respect just like the earlier and continuing uprising of peasant communities against the old forest authority—took a heavy toll on the existing ranks of forest experts, most of whom were members of the Foresters' Union (or else looked to it for protection). The above-noted reorganization of forest affairs in Kherson province, for example, had resulted also in the "total

---

[32] Georgii Vysotskii, "Neskol'ko slov o lesopole: 1. Udobstva natsional'izatsii," *Lesnoi zhurnal* 1-2 (1918): 38.

[33] Ibid., 38-39.

[34] V. M. Bortkevich, "Moskovskii oblastnoi s"ezd soiuza lesvodov," 115-16.

elimination of the local . . . forest specialists."[35] Not surprisingly, the Union itself began in many cases to see the new regime as just as much of a threat as hostile locals. From this perspective, at the above-noted Moscow Regional Congress of the Foresters' Union in January 1918, a "small group" of forest personnel came forward to express "not simply a negative attitude toward Soviet power, but a refusal to work with it altogether, even to the point of [threatening to] go on strike if [Soviet] members should appear in [forest] establishments." In the ensuing debate, a second group suggested as a compromise, that foresters might instead "agree to work for that order which is already *de facto* established in the provinces, that is, cooperate with representatives of the new government but only so long as [the latter] did not interfere with the internal order of their work." (In fact, and as the quotation immediately preceding confirms, Soviet power was at this point far from being "*de facto* established" except in a relatively few specific areas.) In part at the urging of V. I. Stankevich, the Congress's chair, it was resolved to pursue exactly this second, cautiously cooperative path toward "working relations" with the Soviets.[36] "We, forest specialists," Stankevich asserted colorfully, "must defend the forest, must not leave it under any circumstances; [we] must . . . function in the forest like the captain on his ship during a storm: he either goes down with the ship, or at the very least, is the last to leave it."[37] In fact, these heroic intentions, thereafter adopted by the Congress as an official resolution, were doomed from the start to failure—for the new forest administration and the larger Bolshevik order rising around it had already committed to relieving Stankevich, the Foresters' Union, and specialists in general of command of their "ship." Politicians, not technicians were to be the new helmsmen.

Here then, was a second body blow to the prestige and influence of Russia's existing corpus of forest experts. Having already suffered by now nearly a year of violence and abuse from provincial populations, specialists now found themselves targeted also by the political center, which planned to subject them to a simple but devastating demotion—from autonomous and influential authorities to subordinate technical organs under the direct executive control of the new

35 Ibid.

36 Ibid., 118.

37 Ibid., 114.

political authorities. In the spring of 1918 an article in *Lesa respubliki* outlined the intended new status of specialists: they were to enjoy only a kind of consultancy role; and their area of competence was to be redefined to cover only "technical functions," leaving "administration" of the forests to the political authorities and population—typically conceived, or at least publicized, as synonymous—via their Forest Sections.[38] There was, of course, significant overlap of personnel between the new central forest apparatus and the older cadres of experts, as well as considerable jostling and turnover. A single individual might function at times in both capacities. These conditions make exact identification and juxtaposition of "specialists" and "politicians" highly problematical, and the terms should thus be considered approximations. Nonetheless, what *can* be said with fair certainty is that despite having managed to develop something of a consensus on matters of forest organization during the 1900s and 1910s the community of foresters was to begin increasingly to fragment and bicker after October 1917. Supporters of the Revolution and of the ascendancy of politicians over specialists began more and more to bump up against those who were dubious of or hostile to the Bolsheviks, and in particular against specialists or anyone who advocated in favor of the latter's autonomy (in any measure). The Foresters' Union, especially, now began a spectacular decline from foresters' chief organization to pariah group. Its explicit support for the Provisional Government in the spring of 1917 and its status as an organ of technical experts, most of whom were unwilling to defer to political experts in strictly forest-related matters, made this so. Throughout 1918 and thereafter, it was to suffer frequent insults in the Bolshevik press,[39] and on several occasions to

---

[38] N. I. Faleev, "Izmenenie polozheniia pravitel'stvennykh lesnykh spetsialistov," 165-67.

[39] The new Forest Department, a typical comment opined in April 1918, always "got its hands dirty" while the leadership of the Foresters' Union occupied itself with "paper work." *Lesa respubliki* 1 (1918): 7-8.

be explicitly accused of "counter-revolutionary" inclinations or activities.[40] Simultaneously, its opinions began more and more to be ignored or slighted, especially in the pages of *Lesa respubliki* (the Bolshevik forest journal). And more or less from the moment the Bolsheviks took power, Foresters' Union candidates also failed generally to win positions in Forest Sections.[41] (The Union was dissolved in 1921.) On the other hand, the Union may have managed — through sheer perseverence — to retain critical influence in connection with the creation of the "Basic Law On Forests," though this is not entirely clear (see chapter 5). All this, in any case, had the effect of making forest politics, which had during most of 1917 been dominated by tensions between the proponents and opponents of a single, scientifically-managed national forest, more clearly triangular in nature. Pitted against each other during 1918 in particular were "political" centralizers (the emerging Bolshevik forest department), "technical" centralizers (represented chiefly by the Council of the Foresters' Union), and decentralizers (local populations, SRs). As the Foresters' Union declined over the

[40] In the early spring of 1918, for example, the Union was explicitly accused by the chair of the Ural Regional Soviet of Deputies (*Ural'skii Oblastnyi Sovdep*) of being a "counter-revolutionary" organization, a charge not confined to this one instance. Within the Ural region, and on this basis, Union members were declared banned from any and all meetings or sessions of revolutionary organs; and any documentation signed by any Union member was declared null and void. Significantly, however, when pressed to provide more detailed evidence of the nature of this counter-revolutionary activity (by Union representatives), no reply was received. See *Lesa respubliki* 2 (1918): 87-88. Hostility against Union members (sometimes referred to derisively simply as "experts") was in evidence in 1918 also at a number of congresses and meetings on forest-related topics, especially related to the drafting of the "Basic Law On Forests." At the Congress of Forest Commissars held in Moscow in late May 1918, for example, Left-SR participants refused to give to representatives of the Union the right to a "decisive voice" in the Congress's deliberations, pointing out that they were "not elected" by anyone. It was noted that all Union members then present, since they were "technicians," were "in our [political representatives'] service." See, "S'ezd lesnykh komissarov." Stenogram, *Lesa respubliki* (1918): 526.

[41] This was reported in *Lesa respubliki* 2 (1918): 77. I have seen no other confirmation of it, however.

coming months and years, however, "technical" centralizers became less visible, leaving once again (perhaps by as early as 1919) a basically two-sided struggle between the central state and its rural populations. Though the fortunes of technical experts improved slightly during the 1920s, or at least stabilized, the new overall dynamic remained in place.

The new forest administration first weighed in against the existing legal order of foresters—and against their plans to continue organizing forest protection in particular—in February 1918. On the 23rd of that month a notice was dispatched from the *Kollegiia Zemledeliia* (effective the previous 1 January) abolishing the old Forest Protection Committees and transferring all their authority, projects, and inventory either to the newly-created Forest Sections within Provincial Land Committees or to other specified local organs.[42] Especially galling and clearly devastating to the cause of traditional forest personnel were the mass dismissals this was to require. Fired immediately were the middle and lower ranks—"the secretary, forest technicians and service personnel of [each Forest Preservation] Committee." "Specialists,"—but no others—the document noted, would be "reassigned to posts in local forest administration . . . ."[43] Similarly, an official report issued in April 1918 noted that henceforth all categories of technicians still working in the forests were to be considered government employees "subordinate" in "administrative relations . . . to provincial soviets and in technical [relations] to their local Forest Section and to the Central Forest Department."[44] Reassignment in reality typically involved the already-noted demotion. In many cases specialist recalcitrance, among other factors, seems to have meant that reassignment did not in fact take place. In others, local animosities meant that reassignments which did occur were not accepted in the field.

The Institute of the Forest Guard and Inspectors [*Institut Lesnoi Strazhi i ob'ezdchikov*]—which had overseen and staffed basic forest guarding for many years—was also abolished that spring.[45] Its responsibilities were transferred "to the people themselves in the form of organs of Soviet power, which, whenever it

[42] GARF. Fond 5520, op. 2, del. 16, st. 14.

[43] Ibid.

[44] "Poriadok naznacheniia lesotekhnicheskogo personala," *Lesa respubliki* 2 (1918): 114.

[45] Apparently by the decree "On Local Administration of the Forests of the RSFSR" (*O mestnom upravlenii lesami RSFSR*).

is required to create a guard for a given forest, will choose "forest watchers" (*lesnye smotritely*) from among fully literate and conscious citizens for direct guarding of the forest."[46]

Official sanction for removal of experts and appointment of new men was, of course, not needed by local populations, which had already been hard at work in this direction for a full year. On the other hand, state efforts to reduce, replace, and demote synthesized easily with still-simmering popular angers, making it almost "open-season" on any forest personnel—from elite academic foresters to lowly forest guards—associated or associable with the old regime. Local decrees, firings, reorganizations, and so on occurred with dizzying speed and frequency throughout 1918, and freely mixed the new political language with old local hostilities. The Executive Committee of the local Soviet in Kirillovskii *uezd*, for example, "reckoning with the unbending will of the working people," early in 1918 eradicated within its jurisdiction the offices of local State Forest Specialist (*kazennyi lesnichii*), and of the forest guard overall. All relevant personnel, specialists or not, were immediately dismissed. Specialists only—and among these only those who accepted soviet authority and "wished to work for the benefit of the working people"[47]—were invited as usual to accept reassignment "in the service of the forest economy" as "employees" under the control of local state organs.

Waves of similar dismissals quickly brought howls of protest and outrage from foresters in various parts of the country, and especially from their Union. Among the general discontent three basic complaints were voiced repeatedly: many forests were being left unguarded altogether; dismissals were being carried out unfairly, even violently; and replacement personnel appointed by local organs paid attention only to local affairs. In the first connection, for example, Members of the Foresters' Union's Vel'sk section (Vologda province) reported in June that "effective 1 January, the entire forest guard [of a local *lesnichestvo*] with a few isolated exceptions has been dismissed from service, with loss of their maintenance, by order of a Congress of Peasants' Deputies, which met toward the end of March. Consequently, the forest has remained without protection [since]

---

[46] *Lesa respubliki* 1 (1918): 35. In fact, Soviet authorities created a new forest guard of their own soon after. See chapters 6 and 7.

[47] *Lesa respubliki* 1 (1918): 34.

. . . ."[48] And this, it was stressed, had happened at a time when the forests were suffering "massive felling" at the hands of the local population. In one local *lesnichestvo* alone (Chadromskoe) "up to 100,000 logs of 10 - 15 *arshins* [about 23-35 feet] in length and from 5 to 8 *vershins* [about 9-14 inches] in width . . ." had recently been gathered, "in connection with which entire [forest] blocks [*kvadraty*] had been totally cleared."[49] Developments in the Vel'sk area were reported also in the pages of *Lesa respubliki*. The journal, despite its growing animosity toward the Foresters' Union itself, concurred in criticizing the dismissals since the local Land Committee—which had carried out at least some of the dismissals and then taken control of parts of the *lesnichestvo*—had subsequently ignored the national good by selling forest land to local *artely*, freely allowing cutting in these areas and paying no heed at all to any "planned economy."[50]

That the dismissals were not only untimely and unwise, but also carried out unfairly is illustrated by dozens of examples, including some already given. The Eniseisk provincial section of the Union (in Siberia near Krasnoyarsk), for example, wrote its parent council in Petrograd on 19 March, 1918 to complain about the dismissal of a local forester and two inspectors (*ob'ezdchiki*) and the seizure of these persons' books, documents, and of some state property by the local Agricultural Committee (*Sel'skyi Komitet*). One of the inspectors, a certain Kosagov, had, according to the letter, also been requested to "vacate his apartment in the state house [*kazennyi dom*]." When Kosagov, "whose family numbers eight persons, was unable to clear out [quickly enough], he was taken by local soldiers . . . and beaten." Two further inspectors from the same area, the letter continued, had also been "threatened and forbidden to go about their rounds checking the forests." Reports from the field confirmed that here too "in connection with the elimination of service personnel" the cutting of the forests had reached "catastrophic proportions, threatening certain . . . areas with complete destruction in the near future." The soviet thus requested that a message be sent to the Joint Executive Committee of the Soviet (*Obedinennyi ispolnitel'nyi komitet sovdepa*) asking them to take "decisive measures to stop violence against forest department personnel and forest guards and toward the restoration of persons illegally expelled." The letter ended on a combative, if vague, note, suggesting

---

48 GARF. Fond 5520, op. 2, del. 2, st. 75.

49 Ibid.

50 *Lesa respubliki* 2 (1918): 114.

that if "normal" conditions of work could not be restored and violence similar to that just described were to occur again, the local Union would be forced to "resort to all means of defense of the interests and rights of its comrades."[51]

Finally, a great many complaints were made by forest personnel that the new guards established in their stead, far from providing viable protection of the forests (as the latter typically claimed) at best possessed only the most rudimentary understanding of forest protection and cared only for short term, local interests. At worst, which seems most often to have been the case, the new guards themselves either directly engaged in the general plunder or else turned a blind eye to it. Although many forests languished without any organized guard at all, in many cases dismissed forest personnel were replaced, either immediately or later, by locally-selected men. Typically, however, these new "guards'" priorities lay elsewhere than in the management and protection of Russia's forests for the use and benefit of the country as a whole. The central council of the Foresters' Union noted in June 1918, for example, that the emerging methods of hiring forest guards "in some places . . . directly by some or other village or region" were "one-sided" and "not always appropriate, since, frequently, the guard chosen in this manner does not fulfill its purpose." In particular the new guards and other forest personnel, typically chosen from among the local peasant population, operated, in the words of the Vel'sk branch of the Foresters' Union, from "an exclusively local point of view, according to which the local population [is the] absolute and unlimited proprietor of the forests. The notion of the forests as a "general state inheritance" (*obshchegosudarstvennoe dostoianie*) is entirely foreign to them."[52] Their priority, accordingly, was to cut. In some cases, dismissed forest guards in this region had been rehired, it was noted, but "only those who had agreed to give up to the local agricultural organizations up to three-fifths of their pay."[53]

The dismissals, reappointments, and transfers enacted here were further discussed and condemned at the June 1918 Congress of the Foresters' Union. The present situation in the forest economy, one of the Congress's resolutions read, "demands . . . protection of the organization of the Forest Guard, which, in

---

[51] GARF. Fond 5520, op. 2, del. 21, st.st. 21-22.

[52] Ibid., op. 2, del. 2, st. 75.

[53] Ibid., st. 76.

addition to defending the forest, plays a large role in carrying out complex forest-technical work." The Congress resolved the necessity of preserving the forest guard overall and of keeping existing guard positions unchanged. Finally, it asserted somewhat sternly, "All those [guards] dismissed arbitrarily, without sufficient basis must be put back to their former posts." And among the currently-serving forest guard, none should be dismissed from their position without the knowledge of the existing local forest apparatus.[54] These demands were addressed to Forest Sections in provincial and *uezd*-level Soviets.

In all, reports came in to the June Congress from *lesnichestva* all across European Russia: from local branches of *the Foresters' Union* in Tambov, Kazan, Penzensk, Nizhegorod, Orlov, Samara, Vologda, Saratov, Simbirsk, and Petrograd. In each case a virtually identical description of local circumstances was given, forcing the following general conclusion: "forest affairs across the whole expanse of Russia are in the same situation[:] unauthorized, plunderous forest cutting is going on everywhere[;] the forest administration, from head forester to [simple] forest worker is being displaced, [accompanied by] violence . . . and not infrequently the theft of forest workers' property."[55]

The new central forest authority appears to have viewed these developments with a mixture of anguish and satisfaction. Men writing in the Bolshevik forestry journal *Lesa respubliki* in particular—including the increasingly pro-Bolshevik head forester, N. I. Faleev—sometimes allowed themselves a few laughs and snide remarks to the effect that "bourgeois" specialists were simply reaping as they'd sown, a reference to their putative traditional "aloofness" and elitism, and that their misfortunes were thus simply the righteous judgement of a rural proletariat enraged and vengeful after decades of abuse.[56]

On the other hand, it could not be ignored that for the most part, rural anger

---

[54] Ibid., op. 2, del. 1, st. 36.

[55] Ibid., op. 2, del. 2, st. 88.

[56] Faleev, for example, made the following remarks: "You [forest specialists] have made your own fate. You yourselves in your own time, burdened with families and needs, without a murmur, and at times with sheer delight, supported the [old] regime. Now you are beginning to understand the natural animosity that has been nurtured within the population for centuries. Now the tables have turned—earlier [they] . . . were indignant when they saw your actions, now you are indignant seeing the actions of the masses." *Lesa respubliki* 1 (1918): 16.

evidenced little distinction between old and new experts, or between tsarist-era and Bolshevik forest work, planning, or legislation. Though Faleev and some others may have seen a measure of justice in the rural masses' antipathy to "bourgeois" forest specialists, no lesser a figure than Lenin himself saw cause for concern. Though Lenin was in a sense the chief architect of efforts to subordinate forest (and other) specialists to politicians, he had in mind exactly that—subordination, *not* ruin. Suggesting a lack of understanding of rural sentiment in general, Lenin as early as December 1917 had apparently "expressed his surprise" at the general "mistrustful attitude" shown around the country toward specialists[57] A decree issued over his signature the same month, entitled "On the Preservation of Groups of Foresters" (*O sokhranenii Sovetov Lesnikov*) criticized spontaneous measures against forest technicians.[58] When this had no effect, a longer circular was issued on 5 April, 1918 by the Soviet of People's Commissars. Also signed by Lenin, it was written "in response to information received regarding massive dismissals of forest specialists" and addressed to "all Soviets of workers, peasants and soldiers' deputies." It offered rhetorical support to beleaguered foresters everywhere, lauding the fact that "since the moment of the revolution forest specialists have not quit their posts [in fact many had] or ended work [but have] continued to provide a connection between the localities and the center, thus enabling the state forest economy to function." "Forest specialists," the communication continued, were generally in short supply across the country and "can not be swapped for other personnel without damage to the forests and thus to the people in general: the forest economy demands technical knowledge." "General dismissal of forest specialists deprives the forest department of experienced and valuable workers, and [renders] the government incapable of putting into practice all the demands of the people." Thus while it was certainly permissible to "lodge complaints with *Narkomzem* [the chief Soviet agricultural organ] about all undesirable persons," it was quite another to "swap technicians, sent out officially to render technical assistance to local organs of Soviet power." The circular went on to reiterate the substance of numerous predecessors in hopes of inculcating a sense of the center's exclusive authority in these kinds of important matters: "The right of final dismissal of forest specialists and their

[57] "Vserossiiskii s"ezd zemel'nykh otdelov, komitetov bednoty i kommun. Protokoly," *Lesa respubliki* 17-19 (1918-1919): 943.

[58] *Sobranie uzakonenii i rasporiazhenii rabochekrestianskogo pravitel'stva*, Dec. 6 (1917), 93.

transfer from one provincial soviet to another belongs exclusively to the central forest administration."[59]

Lesa respubliki quickly praised the circular and expressed the hope that "the localities will appreciate the significance of this declaration." On the other hand, perhaps concerned lest the relatively pro-specialist tone of the circular be "misinterpreted" as a defense of the Foresters' Union (which it probably was not, in any case)—and perhaps also so as not to have Lenin's words seem to undermine the journal's own insults of Foresters' Union specialists—the journal added several points of interpretation clearly going beyond anything explicitly written in the declaration itself. Thus, "comrade-foresters" were urged to "understand" its "real" meaning, specifically, that it "had in mind genuine [*nastoiashchye*] forest technicians," those who "in word and deed were completely on the side of soviet power [and] for whom serving peasant and worker affairs constituted not a way of getting paid but the genuine work of their whole life." The declaration, it was stressed, could not be used as a cover by those not wholly committed to the soviet system.[60] Taken in context these comments seem perhaps indicative of efforts by the new forest administration, especially in fact by N. I. Faleev, who was also editor of *Lesa respubliki*, to achieve status within the new order, to become part of it, but at the same time to retain for himself some degree of authority and autonomy. Faleev, who went on to pen the Bolshevik "Basic Law On Forests," did not originally and entirely agree with the new political order overall, especially its centralizing bias. Rather, he harbored in forestry some ambitions toward a modest level of decentralization and peasant autonomy. Without ever admitting as much, nor owning up to any modification of his ideas, however, Faleev had by late spring 1918 essentially given up on his localist, "socialization," tendencies and fallen in line with the "nationalizers." This story will be given more attention in chapter 5.

Other Problems

If relations among the three main constituencies identified above—rural populations, the existing corpus of experts (especially as organized within the

---

[59] "Predpisanie vsem Sovetam o nedopustimosti uvol'neniia lesnykh spetsialitsov," *Dekrety Sovetskoi vlasti*, 2: 54-55.

[60] *Lesa respubliki* 2 (1918): 67-68.

Foresters' Union), and the central forest administration—provided the key dynamic shaping Soviet forestry during the early post-Revolutionary period, other factors also played critical roles, especially in matters of forest conservation. The Treaty of Brest Litovsk, for example, signed in March 1918, greatly reduced the area of forest under Russian control (including much of the most easily-exploited and economically important areas), thus diminishing quite dramatically timber reserves available to the state. This occurred, moreover, at a time when timber resources were in particular need as war materials (especially for fuel and railroad construction). In the reckoning of E. E. Kern (a frequent contributor of news of articles about forest politics in the late Imperial, Provisional, and very early Soviet periods), writing in *Lesnoi zhurnal* some months after the signing of the treaty, the loss amounted to "more than 34 million *desiatinas*" (about 37 million hectares), perhaps some 5% of the RSFSR total. Put another way, in signing Brest-Litovsk Russia had, for the time being at least, lost an area of forest nearly twice that of Austria-Hungary's pre-War total, three times that of Germany, four times that of France, and eight times that of Italy.[61] Out of Russia went "almost the entire Dnepr basin, and those of the Visla and western Dvina."[62] The money value of these forests (considered solely as sources of timber and other saleable forest products) was also huge, representing some 40% of the state's total forest profit in 1913 (35 million rubles out of a total of 96 million). The area lost also included many institutions of forest education and training.

If the territorial losses suffered at Brest-Litovsk were soon to be reversed (by November 1918), other factors, also complicating forest affairs, were to persist longer. Especially important was the so-called "fuel crisis" (*toplivy krizis*), which is discussed in more detail in chapters 6-7. The fuel crisis can be described as a relative shortage of timber in an economy unusually highly dependent thereon for fuel (due to the lack or loss of alternatives). "Wood," it was noted in April 1918, was "the only means of powering railroad traffic."[63] It was similarly indispensable in heating and cooking uses, construction, and so on. The fuel crisis seems to have begun during the First World War, but to have escalated steadily thereafter, peaking shortly after the Revolution. Thus, whereas approximately

---

[61] E. E. Kern, "Brestskii dogovor i russkye lesa," *Lesnoi zhurnal* 6-8 (1918): 143.

[62] Ibid.

[63] *Lesa respubliki* 1 (1918): 53.

56% of all fuel purchases in the Russian economy went on wood during 1916,[64] the figure rose to a staggering 85% during 1918-1919,[65] before falling off somewhat to 62% in 1921.[66] Thenceforth, the place of mineral fuels in the economy increased steadily, though as late as the mid-1950s timber remained a major source of fuel especially in northern areas where forests were abundant and transportation costs for coal were significant.[67]

Taken together, the Brest-Litosvk Treaty and the fuel crisis greatly reduced supply at a time of rapidly rising demand.[68] On these two counts at least, not only rural populations, then, but the state itself, was far from ideally disposed to institute and maintain useful levels of protection in the forests remaining. Though careful husbanding of existing resources might at first thought seem to make sense in this context, actual conditions and priorities, as will be seen, appear rather to have encouraged gross over-cutting by the state itself (often to be accounted for as a process of "borrowing" timber from future years' budgets) and the suspension or ignoring (officially or otherwise) of recent and emerging legal protections.

* * *

Overall, then, huge, perhaps overwhelming challenges were already in evidence by the spring of 1918 as efforts to pass a comprehensive forest law approached fruition. Though much of the enormity of the task appears to have been understood at the time, confidence among the new political leadership and the great determination of most forest personnel (both those supportive or skeptical of the Bolsheviks) combined vigorously in search of improvement and progress. As the centerpiece of early Soviet forest management—exploitation and protection—the "Basic Law on Forests," to which we now turn, carried a heavy cargo of hopes.

---

64 RSFSR, *II sessiia Vserossiiskogo Tsentral'nogo Ispolnitel'nogo Komiteta.* VIII sozyva. Vtoroe zasedanie. Stenograficheskii otchet (Moscow: Gosudarstvennoe izdatel'stvo, 1922), 49.

65 Ibid.

66 Ibid. Pervoe zasedanie, 19.

67 L. G. Chertov, "Ekonomicheskaia otsenka lesov Evropeiskogo SSSR," 25.

68 On such matters see, for example, *Lesa respubliki* 2 (1918): 67-70

V
The "Basic Law on Forests"

The "Basic Law on Forests" (hereafter, "On Forests") was signed into law by Lenin and Sverdlov[1] on 30 May, 1918. It remained in legal force (technically, but not always so clearly in practice) until 1923 when it was officially superseded by the "Forest Code" (*Lesnoi kodeks*). It was a complete forest constitution, covering—if often in rather generalized terms—all major aspects of forest organization, exploitation, and protection in Russia.[2] In relevant Soviet histories (especially those published during the "environmental decades" of the 1960s and 1970s) it has enjoyed a positive—if never very critically examined—reputation as a major socialist transformation of forest affairs in general.[3] Regarding

---

[1] Yakov Sverdlov was head of the Soviet Central Executive Committee (*Tsentral'nyi ispolnitel'nyi komitet* or TsIK).

[2] Although the "Basic Law on Forests" pertained specifically to Russia, a similar decree was enacted some two years later—on 3 April, 1920—for the Ukraine. According to an overview of this decree read at the Foresters' Union's 4th Congress in June the same year, the decree differed from its Russian counterpart in only one significant way: article 1 of the Ukrainian law refrained from nationalizing forest areas belonging to peasant communes, an area of forest totaling some 5% of the country's total. Areas within this un-nationalized zone could nonetheless be declared "protected," "water preserving" (*vodookhrannye*) or "monuments of nature" (*pamiatniki prirody*) either "by cash purchase or exchange for other forest areas." The condition of Ukraine's forests—greatly smaller than Russia's—was, the Congress was informed, no better than those in Russia, with extensive areas badly damaged and conservation and regeneration work in general disarray (GARF. Fond, 5520, op. 4, del. 1, st. 5).

[3] See, for example, G. A. Boguslavskii, "Iz istorii okhrany prirody v pervye gody Sovetskoi vlasti." In *Stroitel'stvo Sovetskogo gosudarstva: Sbornik statei k 70-letiiu doktora istoricheskykh nauk, professora E. B. Genkinoi* (Moscow: Izdatel'stvo "Nauka," 1972); N. A. Gladkov, *Okhrana prirody v pervye gody Sovetskoi vlasti* (Moscow: Izdatel'stvo Moskovskogo universiteta, 1972); P. V. Vasil'ev, "V. I. Lenin o lesakh i lesnom khoziaistve." In *Lesokhoziaistvennaia i lesomeliorativnaia nauka v SSSR, 1917-1967* (Moscow: Lesnaia promyshlennost', 1967); Irkutskaia oblastnaia stantsiia iunykh naturalistov, *Vladimir Il'ich Lenin i priroda: Beseda dlia shkol'nikov* (Irkutsk, 1980); I. I. Khutortsov, "V. I. Lenin ob okhrane

conservation in particular, it has been portrayed as a watershed event: it dealt a fatal blow to the preceding era of "forest anarchy," "capitalist predation," and negligible protection; and it laid the foundations for more rational, sustainable, and socially beneficial forest use. That pre-Revolutionary forest protection is not the desert Soviet scholarship has made it out to be has been argued already in chapter 2. In this and the following chapter, it will be asserted that "On Forests" itself brought few real improvements in forest conservation, reducing still further the tangible differences in this regard before and after 1917.

Understandably enough, Soviet literature has also represented "On Forests" as a direct consequence of the Bolshevik Revolution and a purely Bolshevik achievement, rather than (as has been argued here) the culmination under Bolshevik control of plans and programs already fairly well developed and enjoying growing momentum prior to the October Revolution.[4] The law has also been linked indelibly and directly in Soviet accounts to Lenin, whose own efforts, motivated by a personal interest in efficient and benign forest use, were supposedly critical to its creation and passage. In many of the more hagiographic works, "On Forests" becomes simply "Lenin's law" to save the forests, an exaggeration of considerable proportions.[5] (The Bolshevik leader's probable role in "On Forests" and in early Soviet forest conservation in general is assessed in chapter 8.)

If the extremes of Soviet praise regarding "On Forests" are to be avoided, however, the impact of the Bolshevik Revolution on the law should not be excessively discounted, either. It was after all, passed by the Soviet government,

---

prirody" (Maikop: Adygeiskoe otdelenie krasnodarskogo knizhnogo izdatel'stva, 1969); P. I. Zhukov, *Leninskye printsipy okhrany prirody* (Minsk: Gosudarstvennyi komitet soveta ministrov Belorusskoi SSR po okhrane prirody, 1969).

[4] Pre-Revolutionary antecedents are acknowledged in some sources but usually only in passing, and never as in any significant way connected to "On Forests" itself.

[5] I. I. Khutortsov, for example, characterizes "many consequent decisions of the Party and the Soviet government in the area of the forest economy" through the 1960s as simply the "further development and clarification" of the "Leninist prescription" contained in this law (see his "V. I. Lenin ob okhrane prirody," 9). This is misleading on two counts. First, post-Leninist forest legislation diverged from "On Forests" more than this comment suggests. Second, and much more important here, while "On Forests" may have contained "Leninist" elements—such as the stress on science and planning—the law was written by N. I. Faleev, not by Lenin as is implied here (and in many other sources). See chapter 8 for more on this.

not by any other. And at the very least, the Bolsheviks' dismissal of the Provisional Government was necessary for the abolition of private forest property, a core feature of the new law. "On Forests" carries the indelible print of the Bolshevik Revolution in other important ways also. Above all, it established an especially decisive role in forestry for central authority and planning. As already noted, earlier Soviet land-socialization decrees (which had focused primarily on agricultural land) were understood at the time and later to be essentially compromises with the "petty bourgeois and provincial views" of the rural population.[6] Of necessity, they had allowed a relatively high degree of local control, in effect acquiescing in the peasants' existing occupation of land. "On Forests," on the other hand, though it was not entirely without its own concessions to a similar reality in the forests, was by comparison a paragon of centralist control, and in this regard at least, probably ranks as the first major piece of post-Revolutionary land legislation more or less fully expressive of Bolshevik priorities (a condition, incidentally, that makes it worthy of much wider scholarly attention than it has so far received).[7] From the point of view of conservation-minded foresters, themselves also hostile to localism and forest fragmentation, this was of course a generally positive development—though other aspects of the law, in particular its elevation of politicians over specialists, were to please them less.

The Law's History

What appear to be the first concrete steps toward the creation of "On Forests" were taken by the Foresters' Union at its inaugural Congress in Petrograd during April and May 1917 (several months prior to the October

---

[6] The reference here is to the land socialization decrees of October 1917 and February 1918. See P. V. Vasil'ev, "V. I. Lenin o lesakh i lesnom khoziaistve," 17.

[7] Early land socialization decrees, particularly the "Decree on Land," also promoted the "labor principle"—that land belonged to those who worked it (the "principle of the application to land of personal labor" as it was on some occasions described). This principle was absent from "On Forests" which, in search of greater central control promoted forests as the property of all the state, rather than of those who labored in them. See, for example, *Protokoly zasedanii vserossiiskogo tsentral'nogo ispolnitel'nogo komiteta,* 4-go sozyva. Stenograficheskyi otchet (Moscow: Gosudarstvennoe izdatel'stvo, 1920), 105-06.

Revolution, it should be noted).[8] Here, a general framework was adopted based on the earlier-noted rough consensus among foresters. Its main plank was "nationalization of the forests and centralization of their management."[9] The Union's Soviet then worked on a general outline for the legislation, though it was subsequently written up, ultimately in several versions, not by the Union Soviet or a Union commission, but by the central forest authority in the person of N. I. Faleev.[10] Though more or less in line with the Foresters' Union's above-declared priorities (and ultimately with overall Bolshevik goals), Faleev's approach, at least up until mid-1918, flirted just sufficiently with localist concessions to upset committed centralizers (especially in the Union), but nowhere near enough to satisfy popular rural sentiment, putting Faleev in some uncomfortable spots, as will be seen.[11]

Faleev read the first section of his early draft to a meeting of the Soviet of the Foresters' Union in July 1917. It called again for nationalization of the forests, and also for "assessing and satisfying" the timber and other forest product needs of the entire population and economy. The reading apparently "provoked huge disputes" centered on the "technical feasibility" or otherwise of the latter priority. Following a secret vote, however, the draft was nonetheless given majority approval, with the dissenting minority abstaining rather than voting

---

[8] There was also a *Lesnaia Komissiia Glavnogo Zemel'nogo Komiteta* which was involved in forest law debate and drafting during the period from the February Revolution to the promulgation of the "Basic Law." It worked in cooperation with the Forest Department, primarily during the period of the Provisional Government. Apparently, there are "no existing documents" from this. See D. K. Sazhin, "27-fevralia. 27-ianvaria. 27-maia," 844.

[9] GARF. Fond 5520, op. 2, del. 1, st.st. 8-10.

[10] *Protokoly zasedanii vserossiiskogo tsentral'nogo ispolnitel'nogo komiteta, 4-go sozyva. Stenograficheskyi otchet* (Moscow: Gosudarstvennoe izdatel'stvo, 1920), 105. Faleev was assisted in this task at various points by several individuals and commissions. See, for example, N. I. Faleev, ed., *Osnovnoi zakon o lesakh, izdanie neofitsial'noe*, v. A commission, including at least one major figure associated with the Foresters' Union (N. A. Kuznetsov) assisted on "purely technical questions" (ibid., vi).Three Left SRs, Kolegaev, Maiorov and Kostin (all members of *Narkomzem*), created various explanatory notes to the decree (ibid.)

[11] Prior to mid-1918 Faleev typically disdained the word "nationalization" (*natsionalizatsiia*) in favor of "socialization" (*obobshchestvlennie*), for example.

against it.[12] Having been here approved, these ambitious goals were to remain a consistent feature of all subsequent drafts, of the final decree itself and, beyond that, of early Soviet forestry overall.

In August a "special forest commission" was put together by the Foresters' Union to further consider the current draft and make preparations for the forthcoming Constituent Assembly (*Uchreditel'noe Sobranie*), at which it was to be reviewed again. The Union commission, however, enjoyed only a "consultative"—rather than a "decisive"—voice, the latter being reserved to Faleev and his associates at the Forest department.[13] Faleev, in fact, from this point on appears to have become increasingly possessive of the draft, and was sometimes rather defensive in the face of criticisms. Both Faleev and the commission, nonetheless, continued to work on it throughout the late summer.

In September, at the Second Congress of the Foresters' Union, Faleev prepared an update on the law's progress: "Everything was done," he later wrote looking back. "[T]he inventory of demands [for forest materials] was totally worked out, and a draft was ready regarding forest-technical personnel and forest exploitation."[14] But things went somewhat awry at this point. According to Faleev's own account of affairs (which differs in some important respects from that represented by materials found in the Foresters' Union's archive, see below), "news from the localities, the Kornilov Affair,[[15]] and a black mood dashed [all] hopes."[16] Encounters at the Congress with local forest personnel from various parts of the country, Faleev asserted, convinced him then that the earlier consensus—and optimism in general—about the future of forest management and

[12] "Istoriia zakona," *Lesa respubliki* 5-7 (1918): 398.

[13] The Commission included the following Union members: N. A. Kuznetsov (then manager of the Forest Department), D.K. Sazhin, its vice-director, and vice-inspector D. I. Tovstoles. See *Lesa respubliki* 1 (1918): 51.

[14] "Istoriia zakona," *Lesa respubliki* 5-7 (1918), 399.

[15] Lavr Georgevich Kornilov, supreme commander of the Russian armed forces, tried in the late summer of 1917 to forestall a Bolshevik uprising and coup against the Provisional government. In the event, his actions were mistrusted by the Provisional government too, especially its leader, Kerensky, who felt Kornilov might take power for himself. Kerensky thus turned against Kornilov, who was relieved of his command and imprisoned.

[16] "Istoriia zakona," *Lesa respubliki* 5-7 (1918): 399.

protection had been lost. People, he noted, were now speaking "in different languages." He noted also the existence of a "wave of reaction"—perhaps against plans to rationalize and plan forest use, but Faleev was not explicit on this.[17] Displeased, he left before reading his draft.

Faleev's walk-out seems to have been provoked at least in part by other considerations, too. Personal and political disputes with other foresters appear to have taken place, for example. And in particular, it is around this very time that one encounters the first suggestions that Faleev, though clearly an enemy of the advocates of a localized forestry,[18] was as already suggested not quite at home among the centralizers either (though he eventually seems to have learned to become so).[19] Specific clauses in at least one (and probably all) of his drafts, for

---

[17] Ibid.

[18] Frequently, Faleev was seen by SRs and other peasant advocates as an opportunist or traitor. He was, moreover, often viewed as part of the same, hated "old regime" system himself, particularly by some SRs. At the Congress of Forest Commissars in May 1918, for example, one Il'inskii (apparently a Left SR) accused Faleev of "deviousness'' (*izvil'istost'*) noting that whereas only a few weeks earlier Faleev had written in *Lesa respubliki* that "Soviets need not reckon with specialists or with the Foresters' Union when making appointments [of local forest workers and managers]," he now had stated "you [local authorities] may not appoint specialists, we will do so from the center" ("S'ezd lesnykh komissarov." Stenogram. *Lesa respubliki* 8-9 [1918]: 537-38). Faleev often fell foul also of the Foresters' Union, though in this case, for being too "localist" in his thinking. After an uncomfortable several months of sitting somewhat in between camps, Faleev seems by mid 1918 to have become in forest matters an unequivocal supporter of Bolshevism and centralization. These matters are taken up further elsewhere in the text. In this connection an SR accused Faleev around that time of "changing his convictions and going over with exceptional ease an from SR-ist viewpoint to a Communist one . . . ." "Bor'ba protiv Osnovnogo Zakona," 461.

[19] As already noted, Faleev's own position regarding the authority to be given to local or central control seems to have been somewhat ambiguous, evidencing perhaps a desire to balance the two and leading typically to criticism from both sides. Thus, he could at more or less the same time criticize central state ownership as simply "private property owned by the state" while dismissing as nonsense criticisms that he sought to create a "municipal forest economy" when what he in fact was establishing was a single "general, national [forest] fund," which would never for "an instant ever leave the general [state] plan." See, respectively, N. I. Faleev, "Kapitalisticheskye osnovy lesnogo khoziaistva," 13; and N. I. Faleev, "Otmena lesnoi sobstvennosti," 48). See also, "Istoriia zakona," *Lesa respubliki* 5-7 (1918): 400.

example, established "special forest areas for municipal exploitation." Within specified parameters, and in cases of need, these "municipal" or "local" forests, as they were sometimes referred to, allowed considerable flexibility to local organs to use and otherwise exercise authority over local woodlands. These measures in particular provoked bitter criticisms from members of the Foresters' Union to the effect that Faleev was "striving to create a municipalized forest economy" rather than a truly nationalized one. Faleev, for his part, rigorously denied this, asserting that he was merely making a few necessary concessions to legitimate local needs without sacrificing the coherence of forest management overall.[20] This did not convince all his critics, however, one of whom voiced a concern felt more widely among the Union elite that any concessions at all to local sentiment were dangerous since provincial Soviets were in many cases heavily dominated by locally-oriented individuals, SRs, and some of the worst forest-abusers.[21] The Union as a whole opposed even relatively small flirtations with localism as violating the integrity and viability of a national plan and as undermining any possible "guarantees" of the "principles of forest protection."[22]

Criticisms of this nature were as yet incipient, however. They peaked a few months later when N. Paevskii, another high-ranking Union member and learned forester delivered a stinging condemnation of Faleev's current draft for supposedly violating in the body of the text the principles of nationalization and forest unity stated in "formal, declarative" style on the first page. Particularly upsetting to Paevskii were proposals to create "local forests intended for satisfying fuel, construction . . . [and other] demands . . . of the peasantry," which he and others (probably correctly) saw as a simple division of forests into two, leaving one portion under central and the other under local control.[23] In fact, though all Faleev's drafts gave certain powers to local populations, the particular version at issue here seems to have included extra, localizing concessions themselves wrung begrudgingly from Faleev and the *Kollegia Zemledeliia* by the Main Land Committee (*Glavnyi zemel'nyi sovet*), and which were subsequently

[20] "Istoriia zakona," 400.

[21] N. Paevskii, "[Doklad] Priniato v noch' s 13 na 14 maia [1]918 . . . ." In GARF. Fond 5520, op. 2, del. 16, st. 72, and passim.

[22] Ibid., st. 71

[23] Ibid., st.st 71-72.

dropped (see below). Though Paevskii's criticisms were directed at the draft more than at Faleev himself, Faleev nonetheless seems to have taken them—along with most of the trials and tribulations of his drafts—rather personally. The suggestion that the draft ought to be reviewed again, this time with the participation of all the leading forest societies and organizations, and then "subjected to a whole series of . . . alterations" did not help smooth matters.[24] In any case, Faleev's decision to walk out of the September Congress did him little immediate good, since it simply allowed other papers, and other voices, to dominate the proceedings—in particular, those read by two veteran specialists and unambiguous promoters of central and national planning, N. A. Kuznetsov and D. N. Tovstoles.[25]

In early October Faleev completed his draft legislation in full. Rather than present it openly, however, he appears to have allowed his insecurities to get the better of him and chose to "hide it in the literal sense of the word." Copies, thus, were made available only "confidentially to the closest, most intimate work comrades."[26] Interestingly, these included N. A. Kuznetsov, the author of one of the papers on nationalization whose vision of nationalization had already been held up by the Union Soviet as a superior alternative to Faleev's own. To other co-workers, assigned the task of writing explanatory notes to various parts of the draft to be presented at the forthcoming Constituent Assembly (*Uchreditel'noe Sobranie*), Faleev gave only fragments of the draft, letting none see it all. Work continued thus through the Bolshevik Revolution and to the draft's next

---

[24] Ibid., st. 72.

[25] See n. 13 for further information on these two men. N.A. Kuznetsov, in particular, played a significant, if essentially consultative role in creating the "Basic Law on Forests." Faleev himself later acknowledged Kuznetsov's work and deemed him a "true friend." On the other hand he also cited "differences . . . in their political convictions" ("Istoriia zakona," 404). The two thus cannot easily be set on either side of a "politicians versus specialists" divide, despite the existence at times of this type of tension between them. As editor of the Bolshevik forest journal *Lesa respubliki*, Faleev, furthermore, would later publish several articles and commentaries by Kuznetsov. By late spring 1918 they appear clearly as allies in support of "On Forests." Some degree of evolution in the views of both, but especially of Faleev's (as noted), seems to have taken place by then. This, in turn, reflects the narrowing of possibilities as official forestry, and the wider political environment, began increasingly to assume definite shape. The papers by Kuznetsov and Tovstoles, incidentally, are reproduced in *Lesnoi zhurnal* 1-2 (1918).

[26] "Istoriia zakona," 399.

appearance at a meeting of the Main Land Committee's Forest Section (*lesnaia sektsiia*, not to be confused with the Forest Sections—*lesnye otdely*—noted several times in previous chapters[27]) during November 1917. This seems to have been relatively uneventful.

By January 1918, just ahead of being debated at the Third All-Russian Peasant Congress, the draft was printed and several hundred copies distributed to participants of that meeting's Forest Section (again, *Lesnaia sektsiia,* not *lesnoi otdel*).[28] Given the peasant and SR make-up of the Congress, it was not surprising that Faleev's draft ran into trouble—not this time for making too many concessions to local interests, but for making too few. Socialist Revolutionaries[29] from Arkhangel, for example, voiced opposition to the draft because its emphasis on national planning threatened to eradicate the advantages their region and its inhabitants derived from naturally-abundant forest resources (which under the draft's provisions would be used to offset the paucity of forest in other parts of the country).[30] Ironically, a representative from Khar'kov province, exactly the type of forest-poor area the Arkhangel speakers had in mind, himself also condemned any plans for forest centralization since it would run counter to his constituency's plans for "local exploitation" of several forests in the area already declared protected by virtue of being located on erosion-sensitive land.[31] If these were strong criticisms, however, they appear not to have been widespread enough (yet) to prevent the draft—and its blueprint for a more or less centrally-planned forest economy—from being accepted without major alterations.

The draft was discussed yet again during February (beginning the 21st) at sessions of the earlier-noted Forest Commission (*Lesnaia Komissiia*), which included, as well as Kuznetsov and some other Union figures, members of the Main Land Committee. This SR-dominated body attached to the *Kollegiia zemledeliia* had been formed primarily to assist *Narkomzem* implement the "Basic

---

[27] Unfortunately, both these words translate into English as "section."

[28] "Istoriia zakona," 400. The "several hundred copies" noted here are listed in a later edition of *Lesa respubliki* (8-9 [1918]) as "two hundred copies."

[29] SR critics of "On Forests" are usually referred to in stenographic records as "left SRs." In secondary accounts in *Lesa respubliki*, often written by Faleev himself, they are sometimes designated "right SRs." It is not clear if the same or distinct groupings are being referred to.

[30] "Istoriia zakona," 400-01; and "Bor'ba protiv Osnovnogo Zakona," 461.

[31] "Bor'ba protiv Osnovnogo Zakona," 462.

Law on the Socialization of Land," and was suitably agriculturally-oriented (rather than forest-oriented) in its thinking.[32] It was on the third day of discussions, and at the insistence of the Main Land Committee, now more vocally than ever in full opposition to the establishment of a single, centralized forest economy, that Faleev's modest "municipalization" provisions were dropped as inadequate local concessions and the "single [national] forest unit" of the current draft split into two basic categories—forests of "local use," to be put under exclusive local control, and forests of "general state use," to function under central control.[33] For the moment, then, a wave of decentralizing sentiment seemed ascendent; and Faleev's modest concessions to it had been superseded by a more sweeping giveaway. An SR "representative" from Khar'kov offered one of several expressions of the anger of the provincial populations, now more than ever in opposition to central forest planning: "The Tsar-Liberator [a reference to Alexander II, who had emancipated Russia's serfs back in 1861] gave the land to the peasants, and you socialists want to take the forest away from the peasants."[34] Other similar comments were offered, also. The draft, thus amended, was

[32] The Main Land Committee here and elsewhere championed the quasi- anarchist slogan "Land and Freedom; Forests and Freedom" (*zemlia i volia; les i volia*).

[33] Istoriia zakona," 402; "Bor'ba protiv Osnovnogo Zakona," 463. There does not seem at this point to have been any specific idea of how much forest would fall into each category. The final decree "On Forests" itself, enacted on 30 May, 1918, also allowed under certain circumstances for exclusion from the Republic's general forest holdings—and thus from the reach of central planning—of certain "minor" forests of local agricultural designation located on "basic peasant-agricultural plots" (usually referred to as "local forests") and in certain other specified instances also (article7). This was a much less significant concession than that being discussed here, however, which resembles much more closely the split made in Russia's forests by the 1923 "Forest Code" (see chapter 7). The "local" or "municipal" forests envisioned in the final version of "On Forests" (but not those under discussion here), appear in any case to have been potential rather than actual, since absolutely all forests were to be designated "general state forests" at first, with local or municipal forests to be established, according to one estimate, about 1 year after the decree's issuance. See *Protokoly zasedanii vserossiiskogo tsentral'nogo ispolnitel'nogo komiteta,* 4-go sozyva. Stenograficheskyi otchet (Moscow: Gosudarstvennoe izdatel'stvo, 1920), 107. In fact, their establishment appears to have been even slower than this, though in many cases it was also academic, since local populations often ignored central directives and used forests more or less as they wished in any case.

[34] "Bor'ba protiv Osnovnogo Zakona," 463.

approved the next month (March) at a plenary session of the Main Land Committee and was then submitted to the VTsIK.

Partly in consequence of the transfer during March of the Forest Department (along with the seat of government in general) from Petrograd to Moscow,[35] the Foresters' Union (and its commission) apparently did not participate in these later discussions of Faleev's draft. The Union's leadership had, furthermore, by now clearly begun to distance itself from a document in which, it felt, the "principles of nationalization or centralization" it had embraced at its September 1917 Congress had been "undermined."[36] The current, relatively pro-peasant draft, then, pleased neither Faleev, the Forest Department, nor the Union.

Perhaps predictably, however, the central-local pendulum swung back somewhat toward centralization once the draft was taken up again on 11 April, 1918 by the VTsIK. While its general outline, having "already been looked over by the leading Soviet parties, the Bolsheviks and the Left S. R.'s,"[37] seems to have excited little debate, the Main Land Committee's division of forests into local and general state holdings was noted, criticized, and passed over for consideration to yet another commission, this one comprising three Bolsheviks—Sereda[38], Rosen, and Vasil'ev—and two Left S. R.'s—Kolegaev[39] and Ustinov. The commission voted down any two-way division of forests, which should have

---

[35] The move served also as an opportunity to dismiss some 120 out of a total of 150 forest department personnel, supposedly on the grounds that this majority were incapable of or unwilling to throw their total support behind Soviet power (*Lesa respubliki* 1 [1918]: 24).

[36] GARF. Fond 5520, op. 2, del. 1, st. 8-10. Unfortunately, no clear indication is given of exactly how the draft failed to promote these ideals.

[37] *Lesa respubliki* 2 (1918): 85.

[38] Almost certainly S. P. Sereda (1871-1933). A Party member since 1903, Sereda was People's Commissar of Agriculture from 1918-1921.

[39] Andrei Lukich Kolegaev (1887-1937) was an important figure in the early organization of the Left SRs. He represented this party in the SNK (see n. 45) of the RSFSR in the capacity of agricultural commissar from December 1917 to March 1918. Following the Left SRs' rebellion against the Bolsheviks during 1918 he switched allegiance late that year to the Bolsheviks, where he worked in supply and transport organs.

ended then and there any remaining hopes for such an approach in law.[40] In fact, the decision simply helped compound the split already well underway between centralizers and their opponents. Thus, ignoring the re-amended draft, the Main Land Committee simply published its own version in the peasant paper, *Voice of the Working Peasantry* (*Golos Trudovogo Krest'ianstva*).[41] In search of legitimacy, however, they credited their document to Faleev (which was technically a near-truth, since most of the text was in fact his), and noted in the accompanying commentary that it had been debated and approved by representatives of the Main Land Committee and other "forest specialists" (which was also technically true). This shadow draft circulated elsewhere also, and for a short time caused significant confusion and friction among interested parties, some of whom—primarily SRs—continued to support it up to and beyond the passage of the official decree itself on 30 May (which they for a short time attacked as an impostor).

Complaints against the emerging official draft of "On Forests" as a central tyranny were expressed also at a Congress of Forest Commissars (*S"ezd lesnykh komissarov*), convened at Faleev's own request in Moscow from 20-30 May.[42] Each Forest Section (*Lesnoi otdel*) of provincial and regional level soviets was invited to send its head and one co-worker. Members of the central forest apparatus and the Foresters' Union were in attendance also, though not with equal voice.[43] Faleev claims to have intended the meeting as a "working" (*delovoye*) and non-political event aimed at thrashing out basic issues on local-central cooperation, with provincial representatives hopefully providing detailed information about local forest conditions as their contribution to the creation of an overall centralized forest economy.[44] Left SRs—Faleev later claimed—instead used the Congress to voice their standard objections.

After seemingly perfunctory overviews by the VTsIK again on 11 and 21

---

[40] "Istoriia zakona," 403. See also "Bor'ba protiv Osnovnogo Zakona," 463-64.

[41] GARF. Fond 5520, op. 2, del. 1, st.. 11.

[42] Faleev's main presentation here was a four-hour report. See "S"ezd lesnykh komissarov," 527.

[43] For information on the distribution of "deciding voices"—a much-contended issue—see, ibid., 525-28.

[44] Ibid., 519-20. See also N. I. Faleev, "Chto sdelano i chto predstoit' sdelat'," 475

April, the draft was discussed in more depth, this time by the SNK[45] (and thus, perhaps for the first time, by Lenin, who chaired this body) and a commission from the VSNKh on 12 May. It was decided at this point to form yet another commission, this time composed of Faleev, Iu. Larin[46] and P. Stuchki. This was granted authority to approve in the name of the SNK all uncontested points of the law, with the more troublesome issues being passed back for further examination by the SNK itself. The Commission—joined by a number of "forest specialists" (likely, the Union's Kuznetsov among them) and certain members of the Main Land Soviet—officially agreed on all points of the draft during a late-night session on 13-14 May. The project was then confirmed by the SNK the following evening (the 14th) and by the VTsIK two days after that.[47] Two final readings took place before the VTsIK on 27[48] and 30 May.[49] On the latter date the decree was finally signed into law.[50]

Much of the source material for this account of the law's background and passage comes from material published at the time or shortly after the fact in *Lesa respubliki,* the official Bolshevik forest journal under the editorship of none other than N. I. Faleev himself. Faleev, indeed, even wrote some of the articles here used. The story is corroborated in the main by other sources (including stenographic records of debate in the Soviet Central Executive Committee) though with one salient exception. While no mention is made by Faleev or in the

[45] SNK—*Sovet narodnykh kommissarov* or Council of People's Deputies, the cabinet, headed by Lenin.

[46] Pseudonym of Mikhail Zal'manovich Lur'e, an active Menshevik from about 1904 to August 1917, at which time he joined the Bolsheviks. Thereafter, he held posts in numerous committees and commissions on agricultural and other affairs. Larin also played an important role getting Russian Marxists in exile back to Russia after the revolution of February 1917.

[47] *Lesa respubliki.*, 4 (1918): 219.

[48] *Protokoly zasedanii vserossiiskogo tsentral'nogo ispolnitel'nogo* komiteta. 4-go sozyva. Stenograficheskyi otchet (Moscow: Gosudarstvennoe izdatel'stvo, Moskva, 1920), 313-21.

[49] "Istoriia zakona," 403.

[50] At this last *zasedanie* a delegation of Left-SRs was present, including Spiridonova, V. Zitta and others. The delegation tried, apparently without success "to introduce into the law substantial changes." See "S'ezd lesnykh komissarov," 533.

sources noted, the Foresters' Union's archival records[51] claim that the Union itself played a key role in altering the draft at the eleventh hour. Specifically, a "delegation" from their number approached and succeeded in "gaining access . . . to the [relevant] Commission of the VSNKh and to the [Soviet] Central Executive Committee" just as those bodies were in their final deliberations of the draft decree, most likely on either 27 or 30 May. The delegation, moreover, wielding the Union's "moral weight," apparently managed to make "the most substantial [*sushchestvenneishye*] corrections [to the draft decree] in the spirit of the resolutions of the 1st and 2nd Forest Congresses," that is, in the spirit of unadulterated, national, central planning. These corrections, the report concluded, were preserved in the final decree "On Forests" as it was signed into law by Lenin and Sverdlov on 30 May, 1918.[52] A much less detailed but identical restatement of this matter was published around the same time in *Lesnoi zhurnal* (published by the Petrograd Forest Society but discontinued in 1918).[53] Both versions also recount "a comparative analysis" made by the Union at some unspecified point (probably in May 1918) "of the articles of the [draft] legislation [as amended by the Union] and the final and confirmed decree." This, it was claimed, clearly showed the Union's success in altering the draft at the last minute.[54] In contrast, Faleev, writing as always in *Lesa respubliki*, describes the final decree and his own first draft as differing from each other "in no significant way."[55] And an editorial essay, also published in *Lesa respubliki*, asserts that the "law adopted by the VTsIK was . . . precisely in Faleev's wording [*redaktsia*]."[56] On the other hand, Faleev does elsewhere concede some "distinctions [*razlichye*]" between his project and the final decree, though he fails to elaborate (and nowhere does he mention any delegation from the Union).

---

[51] As collected together in GARF, fond 5520.

[52] GARF. Fond 5520, op. 2, del. 1, st.st. 8-10

[53] The Society had very close ties with the Union, and most of its personnel appear to have been high-ranking Union members.

[54] "Vtoroi delegatskii s"ezd Soiuza lesovodov," 337-38. Unfortunately, *Lesnoi zhurnal* did not print the comparative analysis; nor have I been able to locate it in GARF.

[55] "Istoriia zakona," 400.

[56] "Levye s.-r. i lesnoi zakon," *Lesa respubliki*, 8-9 (1918): 485. It is quite likely that Faleev himself wrote this statement.

In the absence of full documentation, including copies of all drafts to compare with the final decree, it is impossible to decide on the veracity of these claims. Nor, in fact, is the matter of great importance other than as a clear example of the tensions between more and less centralizing tendencies and advocates. Further, given that Lenin would have participated in these final key debates, and given also his own commitment to centralization, it seems reasonable that a delegation from the Union would have found a sympathetic ear. Probably the increased centralization it advocated would have been carried out in the SNK even in the absence of the delegation.

"The Basic Law On Forests": An Overview[57]

"On Forests" began by reiterating in a form specific to forests the abolition without compensation of all private property in land already stated in "On Land." Forests were henceforth to constitute the "general people's inheritance" (*obshchenarodnoe dostoianie*) of the RSFSR (Russian Republic) under the general jurisdiction of a revamped central forest apparatus (officially created in January 1918[58]), now titled the Central Administration of the Forests of the Republic (*Tsentral'noe upravlenie lesov respubliki*, hereafter TsULR). (Though destined to undergo additional name changes, this continued to be referred to frequently as the "forest department" or "central forest authority"—a convention that will be followed here.) The TsULR in turn came under the control of the main agricultural organ, *Narkomzem*. In fact, and despite the absence of any explicit reference in the decree itself, neither TsULR nor its parent were destined to have the forests all to themselves. Actual forest exploitation—the physical work of cutting, preparing, and transporting timber (usually referred to as "zagotovka," or procurement) and so on—was to remain largely under the separate jurisdiction of the main economic organ, the VSNKh (specifically, its relevant dependent entities, especially *Glavleskom* [the Chief Forest Committee] which was founded in early 1919 [see chapter 6]). This left TsULR with overall

---

[57] All references to the law in this chapter are from the version printed in *Dekrety sovetskoi vlasti,* vol. 2 (Moscow: Gosudarstvennoe izdatel'stvo politicheskoi literatury, 1959), 312-29.

[58] *Lesnoe khoziaistvo SSSR za 50* let, 16.

administration, planning, and technical work. The resulting overlap was to be a source of considerable tension (see chapters 6 and 7).

The decree established ten chief tasks of soviet forestry, many of which were simple restatements of already-publicized priorities. First listed was the need to "define general technical bases" upon which forestry everywhere in the country would then be required to function "for the purpose of creating a proper [*pravil'noe*] national forest economy and more productive use [of the forests]" (article 9). Other listed goals included: establishing the scale of—and determining norms for—the timber needs both of the national economy and of individual citizens; working out a system for satisfying all these needs; creating local "forest organs" and "technical parties subordinate to the Central Administration;" raising the level "of forest and forest economic knowledge among the population;" and creating a government plan for the exchange of forest and agricultural products, improvements in the transporting of forest materials. Also noted was the priority of "converting forest areas . . . for the increase of agricultural land" (article 9). Conservation, significantly, was not among these general priorities, though several articles elsewhere in the decree are explicitly devoted to it; and the key priorities of planned and efficient forest use, developed in some detail throughout the law, both relied on and sought to integrate conservationist techniques and goals. Overall, however, conservation was here imagined, as it was to be generally in Soviet culture and legislation of the period, as a subordinate concern and a means to an end (efficient exploitation), not as a priority in and of itself.

"On Forests" offered new details about the envisioned forestry hierarchy. From the bottom up: To the village- (*sel'skye*) and *volost'nye*-level forest organs was assigned responsibility for various day-to-day affairs including receipt of declarations from individuals and workers' organizations of their needs for forest materials, evaluation of these and of current disbursements and privileges, issuance of certificates for receipt of materials, and so on. Overall, however, these organs' powers were consultative rather than decisive. Thus, as "organs of Soviet power" they might, for example, make "declaration[s] to the [central] authorities" regarding the need for reforesting a given area or for declaring it "protected" (article 48). But to initiate such processes was beyond their jurisdiction.[59] Likewise, regarding their powers over local forest and technical personnel they might legally "present notices" calling for the rejection of given individuals, even those recently assigned by the center, but they could not actually dismiss or

[59] N. I. Faleev, T. Nekhoroshev, D. Sazhin, "Lesnye organy," 335.

transfer anyone without official sanction (though unsanctioned dismissals and transfers were, in fact, still a common occurence at the time).[60]

The next level up, the *uezdnye* Soviet forest organs, were required to "manage forests situated within the borders of the *uezd*," determine norms for the scale of forest demand and establish quotas for forest material use within their borders, pass forest-tax-generated and other funds onto the center, and issue orders on all types of forest use according to quotas (article 49). In a published article expanding on various aspects of the law, *uezdnye*-level organs were in fact described as the "most important, most significant" of all levels within the new hierarchy and could, in "all matters encompassing provincial, regional, and even state interests," present the center with their opinions if these were also "connected with the interests of the *uezd*."[61] In this connection, the *uezd* appears overall to have been intended as the key local level—and its organs as the key organs—through which local-central communication would operate. Perceptions of these organs as populated with the least bureaucratic and most politically reliable elements seem to have been the chief issue here. Whether or not the greater influence of *uezdnye*-level organs was already apparent at the time of the decree's issuance cannot here be determined; but it was certainly evident to some writers only a few months later. Writing in *Lesa respubliki* in November 1918, for example, one "Tertius" argued that the "center of gravity in forest work" had already begun to shift toward the *uezd*. This, he claimed, was largely attributable to the changed legal and political environment, in which the traditional operations of the provinces (*gubernii*)—which he characterized as essentially political entities, and ones hampered by their excessively "bureaucratic" and "administrative" character—were no longer of such relative importance. "*Uezdy*," he wrote, further defining their specific envisioned role, "will fulfill an economic [as opposed to political] function, and thus their work will become more intensive in the future. The *uezdy* will need to be thought of as economic cells." This obviously accorded well with the emerging division of labor whereby the center monopolized politics and delegated specific economic, technical, and other tasks to local organs. Ultimately, Tertius mused, the future forest economy might well be based entirely on a relationship between the center and "a few *uezdnye*

---

[60] Ibid., 336.

[61] Ibid., 337.

economic cells."[62]

Another possible incentive for promoting the ascendency of *uezdnye*-level organs may have been the recalcitrance of organs at other levels. Regional-level—or *oblastnye*—organs especially appear to have been sources of potent challenges to Soviet forest organization in general and to "On Forests" in particular. The Moscow Regional Forest Section, for example, proposed to the central forest administration in the spring of 1918 that the forestry functions the latter proposed to fulfil should instead be carried out "exclusively through the *oblast'*."[63] More strikingly, around the same time representatives of the Ural *oblast'* ignored "On Forests" and put out their own law on forest "nationalization" based essentially on the more localized approach of the "Basic Law on Land." They backed this move up with an explicit claim to legitimate jurisdiction over regional forest affairs in general.[64] Even the Moscow Regional branch of *Narkomzem*—TsULR's parent—expressed dissatisfaction with the intentions and actions of the center via its own weekly journal *Zemlia*. In the May 21st edition, for example, an article was published accusing the central forest apparatus—TsULR—of "absolutely not wanting to reckon with local organs of authority." In support of this contention the article's author (identified only by the initials V. Z) claimed TsULR was trying to centralize appointments of "absolutely all" forest administration officials—a charge unequivocally rejected by the latter.[65] "V. Z." also condemned the center for supposedly withholding credits or otherwise bringing pressure to bear in cases where centrally-appointed individuals had been refused or dismissed by the localities. Whether in retaliation for these and other regional-level challenges to Soviet power in the forests and elsewhere or for other reasons, "On Forests" in fact gave the *oblast'* no functions at all. In fact, it did not even mention it.

Provincial (*gubernskye*) level organs of Soviet power, on the other hand, enjoyed the same rights within their jurisdiction as their *uezdnye* counterparts did in theirs, plus a few others. Overall, however, their "tasks" were estimated to "boil down chiefly to supervision, management and unification of the activities of

---

[62] Tertius, "Revoliutsiia i mesta," *Lesa respubliki* 16 (1918): 856.

[63] N. I. Faleev, "Chto sdelano i chto predstoit sdelat'," 473.

[64] Ibid.

[65] "Levye s.-r. i lesnoi zakon," 481. Based on evidence cited in various places throughout this study, the accusation in fact seems reasonable.

the *uezdy*."[66] Specific functions assigned to provincial-level organs included "keeping account of forest demand." This was seen as of particular importance since "these data form the basis of general state estimates for national forest demand."[67] Regarding their relations with the central authorities, provincial-level organs could "participate . . . in the resolution of questions about all types of forest use" (article 50),[68] which in fact seems to have meant little.

Suggesting the relatively high priority accorded to conservation, and certainly the centrality of such to the overall envisioned forest plan, local organs at all levels were "obligated to render all types of assistance to conserve forests." These priorities were to be observed not only in connection with forest cutting, but also regarding timber "removal, delivery, transport, floating, and so on" (article 51). Organs at all levels were also to "defend the forests and look after their regeneration," and to use "all measures to increase the areas of forests on the territory under their management . . ." Suggesting again the utilitarian priorities underlying forest conservation at the time, however, these latter responsibilities were to apply only "where this is necessary for achieving the [assigned] quota for forest cover [*lesistost'*]" (article 54).

The obvious major and recurring general theme of "On Forests," however, was centralization. A comment in *Lesa respubliki*—that "On Forests . . .[is] permeated everywhere with a single idea . . . control by the federal authority"[69]—would have been entirely obvious to anyone who had read the decree. The center was made solely responsible for assessing and setting the scale of demand for forest products by all users throughout the country (essentially state and popular needs), and for satisfying this "within the framework of a state plan for the [entire] forest economy" (articles 65).[70] It was to this end required also to assure

---

[66] N. I. Faleev, T. Nekhoroshev, D. Sazhin, "Lesnye organy," 337-38.

[67] Ibid., 338.

[68] Ibid.

[69] "Tsentral'noe Upravlenie Lesov Respubliki," *Lesa respubliki* 5-7 (1918): 343. Ironically, the comment is almost certainly Faleev's (though no writer is credited). Faleev had by now apparently given up advocating for even the slightest amount of localism and become an ardent pro-Bolshevik and centralizer.

[70] Or as restated in *Lesa respubliki*: "In order to bring about communism in reality, the [B]asic [L]aw [On Forests] reserves to the central authorities the right: to determine the need of the individual citizen for timber as a product, to determine the scale of the need for timber by the economy overall, and to

also the "continuous regeneration" of forests, setting conservation at the heart of planning, even if it had not been explicitly mentioned earlier in the decree (article 66). To achieve these goals, upon the TsULR was placed "the responsibility of controlling the actions and directives of all organs of Soviet power, relating to forests on the territory of the Republic" (article 67). This explicitly did not include abolishing any requirements of "On Forests" itself, but did require that the center hold local organs strictly to the "acts and directives" of the decree (articles 68-70).

In fact, even more than this appears to have been envisioned, though it is not explicit in the text of the decree. An explanation of the rights and role of the central forest apparatus, published in *Lesa respubliki* immediately after the law's publication, noted that the TsULR was obliged and empowered by "On Forests" not only to enforce all provisions of the law, but also to act on its own initiative to enforce the *spirit* of the law even when its letter had not been violated. Thus, in the event that "the actions of any local authority, *even though not going against the law* [emphasis added], cause or may cause the forest damage," the "federal authority is obliged to protect this area and stop the harmful action allowed by . . . [any local organ]." The same technically extra-legal reach was to be availed of also "when there is the danger that the interests of the whole country or a part thereof" were at stake.[71] Assertions of this kind, of course, whatever they imply for the improvement of forest conservation, suggest also the limits of the rule of law and the arbitrary authority of the center in the early Soviet state.

Given the ascendance of localist sentiment around the country, the supremacy of the forest center was a concept in need of fairly careful marketing in the very early Soviet period. To push it too aggressively was to risk provoking a backlash from the rural population, the SRs, or others. Excessive timidity, however, would simply mean surrender. *Lesa respubliki*, as the central forest apparatus's chief mouthpiece, resorted at times to both approaches, though in general it appears to have begun cautiously but over a period of a few months following the decree's issuance become increasingly assertive. The former approach was evident, for example, in an article in June which noted that there was "not a single article [in 'On Forests'] which would grant to the federal

---

indicate the method and system for satisfying all these needs. Here is the essence, the sense of forest communism toward which the Basic Law is moving." N. I. Faleev, "Godovshchina revoliutsii," *Lesa respubliki* 16 (1918): 837.

[71] "Tsentral'noe Upravlenie Lesov Respubliki," 346.

authority the right to display autocratic power and not to be under the control of the country at large." Even in "such centralized questions as the establishment of sawmills, which was to be carried out in accordance with a general plan for the state economy, the localities will collaborate, and not a single local organ of Soviet power can be deprived of its right to a voice or be sidestepped." Soviet centralization, the article concluded, in forestry at least, "was not tsarist centralization, but the authority of all the people [*vsenarodnost'*], the authority of all the state [*vsegosudarstvennost'*]." The central forest apparatus was not to be an autocrat "waving the conductor's wand," but an organ whose "entire essence rested on reciprocal, definite, necessary collaboration with the organs of local soviet forest power."[72]

Power sharing of this sort—at least in the realm of "action" if not of "thinking"—was, of course, a genuine desire on the part of many in the new central forest organ. It had been part of the earlier foresters' consensus, and remained an important priority for Faleev personally, who had struggled to include in his centralizing document some degree of local autonomy, as already noted. Beyond this, however, some level of central-local cooperation was also a basic necessity. For one thing, during a time of war and revolution, the establishment of communication between center and localities was far from easy, rendering local initiative vital, if also potentially dangerous. "Life is developing quickly and irrepressibly," noted the pseudonymous "Ne-Lesovodov" writing in this regard in the 15 April edition of *Lesa respubliki*, two weeks before "On Forests" was issued:

> The localities cannot and dare not wait for orders from the center, [not] while matters are urgent and insistently demand intervention and resolution. Regarding the forests one can say that thanks to a whole series of circumstances . . . the connection between the center and the localities has been broken. The mail works slowly, funds don't get where they are assigned. There is no time to sit around and discuss [things]."[73]

---

[72] N. I. Faleev, T. Nekhoroshev, D. Sazhin, "Lesnye organy," 340.

[73] Ne-Lesovodov, "S"ezd lesnykh komissarov," *Lesa respubliki* 3 (1918): 135-36. The state of communication between the center and localities—and an example of the necessity of allowing some freedom of action to the latter—was clearly indicated in a forest department report published in May 1918. This stated that between 3 April and 3 May of that year fifty-four variously-organized and named provincial-level forest establishments (*uchrezhdeniia*) had consulted the department 1,093 times. Participation among these local organs had been highly uneven, though. One in Tobol'sk province topped the list with 73

Faleev offered a similar point, but this time from a geographic perspective, in a book commenting on a late draft of "On Forests": "[G]iven the variety of soil, forest-economic, and technical conditions [governing the] growth of forests, it is necessary to give to . . . legislation broad creative freedom, without excessive constraint." This type of "creative freedom," he added, however, could of course only operate "exclusively within those basic molds [*formy*] which are outlined by the basic forest law."[74] Some level of cooperation with, rather than domination of, the localities was also the only viable option at the time, acknowledging as it did the present relative power and independence of local organs and populations, while clearly striving at the same time to reduce this in the short to medium term.

A much more belligerent tone, however, was evidenced later the same year, in many cases apparently directly contradicting what had been said only recently about the relation of the center and localities. *Lesa respubliki* published the following in October, for example:

> The task of the center and the task of the Basic Law On Forests [is to] centralize the forest economy of the country, to dictate [diktovat'] its will to the localities and demand from them absolute economic obedience, for independence and autonomy of the localities cannot create a socialist distribution of products or satisfaction of all demands. In such circumstances the task of the center may include declaration of a decisive and unwavering war against all separatist forest strivings. A unified economy demands unified measures, coordinated with all others, rigorously thought through and firmly implemented in the interests of satisfying the nation as a whole.[75]

These and the many other various interpretations given of the nature of

---

communications, followed by counterparts in Vologda (65), Novgorod (60), Olonets (59), Viatka (48), and Kostroma (47). Pskov, on the other hand, communicated only once, Vakinsk, Dagestan, Fergana, Erivan and Semipalatinsk each racked up two communications. Several local organs did not communicate with the central forest authorities at all, specifically, from the Don Valley, Urals, Kuban, Tver, Chernomorsk, Zabaikal, Samarkand, and Sakhalin. Citing the existence of a wide variety of reasons behind the silence or near silence of so many regions, the department noted that it had no idea whether or not, or to what degree, either Soviets or their component Forest Sections (*Lesnye otdely*) had been organized there. *Lesa respubliki* 4 (1918): 221-22.

[74] N. I. Faleev, "Programma i sistema osnovnogo zakona o lesakh." In N. I. Faleev, ed., *Osnovnoi zakon o lesakh, izdanie neofitsial'noe* (Moscow: Izdanie Narkomzema, 1918): 33.

[75] *Lesa respubliki* 14-15 (1918): 741-42.

emerging central-local forest dynamics appear to have been essentially "marketing strategies." They did not reflect real diversity of opinion about the basics of the "organic centralism" already described.

On the other hand, there was much (and genuine) ambiguity about the specific forms and functions of the new organs, especially at the central level. An essay outlining relevant matters in *Lesa respubliki* during June noted rather vaguely, for example, that central or "federal" power "will be the power of all the Russian Federation, once everything has settled down, [once] local organs have come into being and become established, [once] their interrelations have been cleared up." "The process of federal amalgamation (*sliianie*)," it was admitted, "is still at its very beginning."[76] Article 72 of the decree in fact explicitly shelved resolving the exact composition and setup of the central forest organ to a separate law to be worked out at a later point. All this, of course, underscores the general dynamic of forestry at the time. Given the ruinous centrifugal forces then running amok, the new authorities were forced to work more or less by trial and error—and always with great caution—toward ultimate goals already laid down in law but for the foreseeable future more desiderata than specific battle plans.

"On Forests" also reaffirmed the ongoing subjugation of specialists to politicians, requiring the TsULR to organize "local forest organs and technical groups" and to make these subordinate to itself (article 9). Article 75 authorized the TsULR to "govern the technical activities of of local organs." D. Sazhin,

---

[76] N. I. Faleev, T. Nekhoroshev, D. Sazhin, "Lesnye organy," 338. Some suggestions for a specific structure for TsULR were offered by N. A. Kuznetsov, speaking as Chair of the Congress of Forest Commissars in Moscow in late May 1918: "At the head of Ts. U. L. R. must be formed a Main Forest Soviet [*Glavnyi lesnoi sovet*] for collegial discussion of questions of the organization of the forest economy. It functions will be: to decide all organizational matters, to discuss projects for instructions regarding the implementation of ["On Forests"], discussion of fundamental complaints against decrees made by local organs, working out rules regarding the service of specialists, and so on. The Main Forest Soviet will meet in session at least twice a year. Its members will be representatives of technical organs of provincial-level Soviet power." See "S'ezd lesnykh komissarov," Stenogram, *Lesa respubliki* 8-9 (1918): 529-30. Above all, this organ had to "establish connections between the center and the localities," to which end Kuznetsov proposed elsewhere that it would comprise "one representative each from provincial Soviets, the Foresters' Union, institutions of higher education, and so on." See "Vtoroi delegatskii s"ezd Soiuza lesovodov," 337.

wrote in *Lesa respubliki* that June that "On Forests" "tears all matters of [forest] management out of the hands of individuals, transferring them to the organs of Soviet power: the law will turn forest technicians into a technical force and compel it to function in this manner as a forest organ." In this regard the *lesnichii*, or chief forester, was targeted in particular. "In previous times the *lesnichii* was closer to the forests [than was the bureaucracy], but he too was [nonetheless still] very distant from people and their needs." Insufficient numbers and the "burden of office work," Sazhin asserted, kept the forester away from his true vocations: the forests and the people.[77] The new law thus in effect codified what had already been done (the dethronement of old foresters as executive authorities or a distinct group of any sort.). The clear distinction drawn in the new law between forest politics and forest technical work was made clear in another article by Faleev.

> The idea of Soviet nationalization [*natsionalizatsiia*] of the forests, the idea of communist centralization, running through every article of the [Basic] law [On Forests] insistently demands that in the local Forest Sections [*Lesnye otdely*] the leading role be taken by forester-specialists [*lesovod-spetsialisty*] in technical matters and by representatives of the communist socialization [*obobshchestvlennie*] of forests in matters of politics.[78]

"On Forests" and Forest Conservation

Forest protection occupied an important position in "On Forests." Though, as noted, it was not explicitly prioritized in the decree, the overarching goal of satisfying efficiently a rising national demand for forest products obviously required better conservation of existing supplies as well as improved forest regeneration for future use. Accordingly, provisions for these were made in several parts of the law, all within the general priority of forest planning. The responsibilities of the TsULR to ensure the "constancy of forest regeneration across the country" [article 66b], and to "define measures" to protect forests from fires and insect infestations has already been noted, for example, and were accompanied by other similar notes throughout the decree.[79] Following forest

---

[77] N. I. Faleev, T. Nekhoroshev, D. Sazhin, "Lesnye organy," 335.

[78] N. I. Faleev, "Chto sdelano i chto predstoit' sdelat'," 475.

[79] Conservation measures were also implied in several other places. Section 5, for example (which deals with "forest economy basics," reasserts that forestry always and everywhere in the RSFSR must be

legislation from the time of Peter the Great forward, the law also provided for a category of "protected forests" (non-protected forests were simultaneously re-designated "exploited forests," articles 77-82[80]).

In addition to reiterating by now traditional bases and reasons for protecting a given area—to prevent erosion, keep waterways open, preserve climate, and so on—"On Forests" offered two new ones: to establish "monuments of nature" (*pamiatniki prirody*), and for "aesthetic and cultural purposes." In fact, aesthetically-motivated protection seems never to have been taken very seriously during the early Soviet period, the concept being considered somewhat "sentimental" and even bourgeois. Cultural and (to a lesser degree) "monumental" protection, on the other hand, appear to have found considerable expression in the early Soviet government's establishment of the series of *zapovedniki* alluded to earlier in this study.[81] These, as already noted, were established primarily for their scientific interest as living laboratories for ecological study, though they served just as well—perhaps in many cases better—as nature preserves (or "monuments of nature") pure and simple. Forest *zapovedniki* were of course to count for only a tiny area of Russia's total forest, making their impact on forest conservation in general essentially negligible. Moreover, they appear also to have suffered from the same kinds of popular-rural violations and abuses that obtained both in other categories of protected forests and across the country more generally.[82] Similarly, forest and other *zapovedniki*

---

carried out in accordance with a "technical plan" (article 79), rather than spontaneously. To this end certain conservation measures would be required: "systematic forest regeneration" and the creation of "protected forests . . . in which use [would be] limited" (articles 77-82).

[80] These were envisioned as sources of "material goods" to "satisfy general-state [*obshchegosudarstvennye*] demands," and provide "monetary benefit" (via the sale of forest materials [article 91]). Their planning was to be organized strictly by the center (article 92).

[81] Not all protected forests were *zapovedniki*. The latter term is a subset of protected forests.

[82] For example, a 1922 survey by the State Committee for the Protection of Monuments of Nature of all *zapovedniki* then organized or in process presents a litany of complaints. In one reserve, there was "mass poaching and the destruction of protected fauna and flora." In another there was "unauthorized cutting of thousands of trees [*stvoly*]." And so on. (GARF. Fond 2307, op. 1, del. 64, st. 64. The first comment here refers to the Caucasus bison (*zubr*) reserve, the second to the Crimean reserve.) A further report, written two years later and focused on affairs within the Vorskla Forest *zapovednik*— specifically

were also violated by the Soviet government, which extracted and sold resources from them on numerous occasions in spite of the technical illegality of such activities. In less direct ways, too, the authorities helped undermine with one hand what they appeared to be establishing with the other. Conservationists, contemplating the problems plaguing their *zapovedniki*, complained frequently and noisily about a general lack of "support"—especially of funds—from the center.[83] This resulted in many cases in the "absence of the most basic equipment," including guards, transportation, signposts, and so on—all of which made peasant violations more or less irremediable while also complicating greatly the transformation of paper plans to real conservation.[84] To some degree, of course, these and other shortcomings are understandable given the context of Revolution and Civil War, the massive dislocations of which would have hindered even the most committed conservationists. Their significance thus should not be unfairly exaggerated.

Even on paper, however, "protected forests" were neither inviolable (in contrast to the *zapovedniki*) nor permanent creations. While they might be organized "with a total ban on all types of use," they might also allow "limited use of timber and products of secondary use" in accordance with an "economic plan" (article 88). The choice of which of these two regimes to implement depended entirely on the overall plan to be implemented and the current needs of the larger economy. Protected status, furthermore, applied only so long as was "necessary" to achieve the "ends" for which it had been granted; and the transfer of protected forests to exploited status could be made by decision of the government level at which the original decision to protect them had been made, or by a higher one (though never by a lower one against the wishes of the center, see articles 86-87). As in tsarist law, then, (and especially like the provisions of

---

since its recent establishment—spoke of whole areas "disturbed and thinned out" (*potrevozhen i razzhizhzhen*), of numerous valuable individual trees cut down or damaged, and of lately increased and destructive cattle-grazing inside the reserve. (GARF. Fond 358, op. 2, del. 8, st. 19.) Writing in 1926 N. A. Kuznetsov cited the same Vorskla Forest *zapovednik* as a prime example of the huge failings of what he called "Russian-style" nature conservation. (N. Kuznetsov, "Ves'ma vazhnyi vopros—pamiatnik prirody—'Les na reke Vorskle'," *Priroda* 5-6 (1926): 97-99.

[83] In 1921 and 1922, at least, there was no funding at all.

[84] GARF. Fond 2307, op. 1, del. 64, st. 64.

the 1888 "Statute") protected forests were not so much untouchable as exploitable on a planned basis.

In addition to listing new rationales for forest protection, "On Forests" also widened the definition of protected forests in general, including for the first time "resort forests" (*kurortnye lesa*) and parks (both areas in which aesthetic and recreational considerations were held to be paramount), as well as forests serving "cultural" purposes, such as those connected to educational institutions.[85]

Regarding protecting forests from fires and insect infestations, the law had little to say beyond a general call for the TsULR to "establish a basis for measures directed toward" battling these and "other disasters" (article 75g) and the imposition on the citizenry of the responsibility to answer "temporary" compulsory labor drafts issued by local soviet organs for the same purpose (article 41). Considering the relatively great importance accorded by the tsarist government to efforts against forest fires and insect infestations (see chapter 2), the pervasiveness and considerable destructiveness of these either side of the Revolution, and the Bolsheviks' avowed aims to run a highly efficient and productive forest economy, one might have expected more.

In addition to the above, "On Forests" also included a seven-article section (articles 29-35) entitled "forest protection" (*okhrana lesov*). This is the most explicit and obvious treatment of conservation in the whole decree. Significantly, however, it is also a sub-section of section 2, entitled "Rights and Obligations of Citizens." In all other sections, conservation is treated as an integral part of the overall plan, rather than as a discrete topic of its own, and is referred to in somewhat random fashion in several connections. The choice to situate the decree's most explicit treatment of forest conservation within a treatment of citizen's rights and obligations may reflect the sentiment among foresters—Bolshevik and not, alike—that the citizenry, rather than industry or the state, was the major immediate threat to forests and the constituency most in need of education and persuasion in this regard. If this was indeed the case, it was obviously rather short sighted.

The section on forest conservation began by establishing the right of all citizens to call to the attention of Soviet organs the need to designate a given forest area "protected," to "limit permissible forest use" or undertake

[85] A. Fok, "Zashchitnye lesa," *Lesa respubliki* 5-7 (1918): 323.

reforestation, and the necessity of any other general forest protection measures (article 29). Interestingly, this meant that each individual in the RSFSR enjoyed the same rights in this regard as did village (*sel'skye*) and *volostnye* organs of government (see above). Whether this is indicative of the high stature accorded citizens in the nascent Soviet state or the paucity thereof accorded local organs is obviously a matter of interpretation. Local Soviet forest organs were empowered to respond to such notices, or at their own initiative to "undertake [whatever] measures they find necessary for the protection of the forests" including the right (following a "technical evaluation") to declare a given forest "protected," or to organize reforestation (article 30). This apparent grant of considerable local power, like all others, was largely illusory. Local measures here as everywhere had to conform to central directives, even if explicit central permission was not needed to do so in each individual case. The center, moreover, could overrule local initiatives whenever it so wished and also reserved unto itself the exclusive power to resolve any disputes that might arise between local organs (article 84).

Article 31 gave all citizens equal rights of access to the forest. This entailed six specific privileges:[86] "to enter the forest"; "to receive fuel"; "to receive building materials"; "to obtain from the forest fallen wood"; "to participate in secondary [forest] use"; and "to make declarations regarding necessary forest protection measures." Though praised in *Lesa respubliki* as a great democratic improvement on old restrictions and inequalities, it did not in reality translate to greater freedom of action in the forests for the peasantry and thus excited much criticism. First, materials could not be "received" or "obtained" voluntarily, but only in accordance with quotas—to be set up ultimately by the central administration as part of an envisioned central plan, but in the short term to be established by local Soviet organs. Quotas, of course, were to be determined not primarily by the availability of timber in a given locality, but by its availability nationally, and by the demands of the national economy, including the needs of industries, railroads, etc. Thus, the residents of especially well-forested areas might find themselves denied the right to cut, or be presented with limited quotas, based on shortages elsewhere in the country. They might have to watch as timber they themselves could not use was shipped off to nearby railroads or other users,

[86] Puzzlingly, the rights of citizens to enter and use the forest are rendered differently—and less liberally—elsewhere in the law. Here, citizens' rights to enter the forest are limited to the purposes of "sojourning [*prebyvaniia*] . . . hunting, bee-keeping, and so on," and "for carrying out generally useful scientific or technical research" (article 31).

or to other parts of the country. Materials they were allotted, moreover — for use as fuel, building materials, and so on — if not used in full within a set time-period were subject to return. Although these provisions made conservation sense overall, they did not enjoy popular approval. Socialist Revolutionary critics of the law expressed common concerns. Said one, in May 1918:

> The citizen . . . had just begun to dream about the beautiful, free spring, when the forest greens and the glades blossom . . . And [now], free from the masters' and the Tsar's lackeys, he will enter his native forest to breathe in the air and listen to the sound of the birds, and will fall into horror, seeing as he enters a sign in the forest: 'Entry into the forest may be permitted only by the Forest Department.'"[87]

The latter charge—and the fear behind it—was no mere paranoia. Article 32 established that "the right of entry into the forest (times, fees if any to be charged, registration, and so on) will be determined by the local organs of Soviet authority . . . ", themselves responsible in such matters to central directives. Citizens and groups might also hunt, keep bees, or carry out "generally useful scientific and technical research" in the forest only on terms and conditions granted by the same organs (articles 33-34). And any income or profit derived from secondary (*pobochnyi*) forest use was to be turned over to local forest organs "for the improvement of the forest economy" (article 35).

Articles 36-42, not explicitly designated as concerning forest conservation, nonetheless outlined citizens' obligations in this very regard. These included "participating in forest-growing [*lesovozrashchenie*] and forest protection (article 36), the specific "character" of which work was to be established by various local organs (article 37). There were also fees due from citizens not participating in forest work (article 38), children's events and work brigades (article 39), and compulsory labor service, all designed to marshal the necessary resources—monetary, physical, psychological—to carry out necessary forest conservation work (articles 40-41). Non-compliance was criminalized (article 42). Though no mention seems to have been made of it, the decree's many critics might also have pointed to its headings, which while assigning "rights and duties" to citizens and local organs, noted in the case of the "central Soviet power" only "rights."

[87] "Levye s.-r. i lesnoi zakon," 487.

Contemporary Assessments of the Law's Conservation Regime

The most positive assessment of the law in general, and of its conservation regime in particular came, obviously enough, from its creator (Faleev) and other boosters at *Lesa respubliki*. An overview of the law's conservation clauses—running to several pages—was printed in the 1 June edition. "Ne-Lesovodov," for example, here lauded it in comparison to the tsarist forest codes as at once kinder, gentler, and more effective. "On Forests", he asserted, did not resort to the "punishments and tortures" and "threats of imprisonment" by which pre-Revolutionary law had sought to protect forests. Instead, "Soviet rule makes no threats; it speaks: [saying] if someone evades his forest responsibilities I will take effective measures."[88] Such measures, he continued somewhat vaguely, could assume the form of "suggestion," "censure," or something "stronger than any fine." In fact, these supposedly "unthreatening" approaches were to be used only against the general population—here construed as "unconscious" violators of conservation law who required not punishment so much as simple "measures to develop consciousness." "Conscious" violators, on the other hand (those class enemies apparently beyond education), would answer to the "implacable, firm, unwavering revolutionary power of all the people"[89] That "all the people" probably actually included very few advocates of "On Forests" seems not to have mattered much here.

"On Forests'" approach to forest conservation also distinguished itself from earlier legislation in Ne-Lesovodov's opinion by focusing not on the symptoms of forest destruction—the unauthorized or inappropriate cutting which had been the focus of tsarist legislation—but on its supposed cause: unequal or unsatisfactory access to forest materials. Proper forest planning was to eradicate these ills and thereby the destruction they had provoked. "If everyone enjoys the right to forest [use], if everyone is satisfied," Ne-Lesovodov suggested, "why would there be any unauthorized cutting? Why would there be any theft in the forests?"[90] In hindsight these might seem rather naive comments—as well as an irresistible tempting of fate. On the other hand, they were also in accordance with

[88] Ne-Lesovodov, "Okhrana lesa," *Lesa respubliki* 5-7 (1918): 321-22.

[89] Ibid., 322.

[90] Ibid., 321.

beliefs that appear to have been widely held by Russian foresters of the day. Among the general positions adopted in April (1918) by a Regional Forest Conference (*Lesnoe oblastnoe soveshchanie*) in Petrograd, to cite just one example, was a statement that the "interests of forest conservation and forest safeguarding" *(lesookhranenie i lesosberezhenie*) demanded at the same time "wide popularization among the population of the idea of proper forest use and the development among the population of a faith that everyone may receive forest [materials] in the quantity actually necessary to him."[91] Suggesting further dimensions to the basic approach and ambition of early Soviet forets preservation policy—and simultaneously finishing up what really must be judged a rather poor exercise in legal writing—Ne-Lesovodov then switched somewhat confusingly from his thus far essentially materialist position to a more idealistic one, concluding that "[i]n our opinion the best means [of protecting forests] is to raise [the level of] education [and] culture, to create such conditions, that every citizen will be conscious, honest and understand his obligations."[92] In other words, here again was a profession of faith in the notion that the ultimate interests of state and society, rural and urban populations, industry and agriculture, were identical.

Somewhat less enthusiastic about "On Forests" as a conservation law was the Foresters' Union—this despite its claims to having significantly influenced the decree at the last moment. Shortly after its proclamation on 30 May, for example, a special commission created by the Union to deal with the decree's impact on the forest guard asserted that the clauses on forest preservation were written "in such generalized expressions" that there was nothing concrete for the union to base its work on."[93] The declining status of specialists, and their frequent replacement in the provinces by relatively inexperienced personnel, also threatened to undermine proper forest conservation in the Union's opinion.

The most biting criticisms, however, came from rural populations and their SR representatives. Conservation, though, appears never to have been among their main concerns (a fact which corroborates the oft-expressed opinion of foresters that rural constituencies cared little for conservation anyway). Rather, attention was focused almost entirely on what was seen as the replacement of one

---

[91] "Iz materialov lesnogo oblastnogo soveshaniia, sostoiavshigosia v Petrograde 16-20 aprelia 1918 goda," *Lesnoi zhurnal* 3-5 (1918): 120.

[92] Ne-Lesovodov, "Okhrana lesa," 322.

[93] GARF. Fond 5520, op. 2, del. 21, st. 125.

central tyranny, that of the tsarist bureaucracy, with another, that of the Bolsheviks. Much the same seems to be true of Menshevik assessments of "On Forests," which also claimed above all that the envisioned central authority over the forests would simply create a new forest "proprietor—the government."[94] All such complaints were countered by the central forest apparatus as errors pure and simple. Mensheviks who had called for the forests to be declared "general-public property" (*obshchenarodnoe dostoianie*)—and then complained that they had instead been constituted as "general *state* property" (*obshchegosudarstvennoe dostoianie*, emphasis added) — were told the two terms were "synonyms" and accused of sophistry. Mensheviks were also charged with grossly misrepresenting the overall nature and intention of forest government in the new Russia, which, in virtual negative image of western models, and, it was implied, of Menshevik intentions too, was "to prevent all actions directed toward . . . private ownership," and to preserve "socialization."[95]

Outside strictly forest-related circles, Russia's leading conservation advocates in general seem hardly even to have noticed "On Forests." For example, A. P. Semenov-Tian-Shanskii, one of Russia's most celebrated advocates of nature preservation during the very late Imperial and Revolutionary periods, writing in the science journal *Priroda* about a year after the decree's issuance, made no mention of it all in a lengthy and impassioned call for Russians to take conservation seriously—except perhaps implicitly in calling "all the decrees so far published" in Russia concerning nature preservation "insufficient."[96] Nor did Semenov-Tian-Shanskii offer any indication at all of seeing the Bolshevik Revolution as a watershed, real or potential, in this area. The conservation movement in Russia, he asserted, began "9 or 10 years" ago with the *zapovednik* plans put forward by I. P. Borodin and others (see chapter 2).[97] Semenov-Tian-Shanskii's only reference to the Revolution at all, in fact, was both indirect and negative. In it he spoke of the need for Russians to understand and appreciate the "purely aesthetic" importance of wild nature, especially "at a time when purely materialistic strivings are at a peak and [when], in spite of the

[94] "Kontrol' federal'noi vlasti," *Lesa respubliki* 5-7 (1918): 343.

[95] Ibid., 344.

[96] A. P. Semenov-Tian-Shanskii, "Svobodnaia priroda, kak velikii zhivoi muzei . . . ," *Priroda*, 4-6 (1919): 209.

[97] Ibid., 206.

great height of the waves of socialist thinking, so too—unfortunately—is egoism."[98]

It is possible that by 1919 the "On Forests" had not yet had enough time to impress itself upon the minds of non-forester conservationists. On the other hand, a well-known article by the Soviet hydraulic engineer, V. E. Timonov, written several years later—in 1922—found abundant recent and ongoing damage to Russia's forests and other natural landscapes and little if any national interest in the question. Unlike in Germany, Switzerland, and a few other places around Europe—where, Timonov felt, protection of forests from the depredations of railroad expansion and so on *was* taken seriously—"In Russia," "there is not even any talk of protecting nature from engineering works."[99] No mention was made here either of "On Forests" or of any forest-related legislation. (The article, however, was focused specifically on the environmental impact of engineering works, not on conservation in general.) Similarly, a note on conservation written early the same year by Gosplan and presented to the All-Russian Central Committee also made no explicit reference to "On Forests." Its assertion that "Russia, in the area of nature conservation [*v dele pokrovitel'stva priorode*] has [always] stood and still stands in last place" suggests yet again either the law's invisibility or ineffectiveness in this regard.[100] Indeed, outside of forestry circles, "On Forests" seems only to have become an important document only in retrospect, well after World War II (see chapter 8).

* * *

Back in the forests themselves, as "On Forests" made its debut in late spring 1918, conditions remained chaotic. Seventeen separate reports from various localities summarized at the Foresters' Union's Second Delegates' Congress during April-May confirmed that the destruction of the forests and dismissal of personnel had not yet even begun to abate. The summary report spoke of

---

[98] Ibid., 213.

[99] V. E. Timonov, "Okhrana prirody pri inzhenernykh rabotakh," *Priroda* 1-2 (1922): 86.

[100] GARF. Fond 2306, op. 1, del. 1988, st. 12. Also cited in G. A. Bogoslavskii, "Iz istorii okhrany prirody v pervye gody sovetskoi vlasti," 267.

> excesses in many places, of major cutting, of individual and mass dismissals of [forest] service personnel from the *lesnichii* to the *lesnik*, of whole groups of forest specialists who had been accused of sabotage simply for issuing a single economic order, and so on. From the localities come petitions for greater support for . . . [local branches of the Foresters Union], for better relations between the center and the localities.[101]

Individual reports from the localities offere d familiar details: In Cherepovetskii *uezd* in Novgorod province "total forest destruction" had occured, along with the dismissal of "virtually the entire service personnel . . . despite assistance [on their behalf] from the central authorities." In the Urals a "new regional [*oblastnoi*] organization, *Uraloles*," had undertaken to "reduce to nothing both the Foresters' Union" whom it accused of "sabotage and counter-revolution" and "all its work . . . ."[102] Overall, communication between the center and localities remained tenuous, cooperation elusive. Exhortative circulars, calling for an end to unauthorized cutting, continued to be written, circulated, and ignored. With "On Forests" completed, the task of narrowing the distance between paper plans and forest realities looked set to begin in earnest. Though tensions had grown considerably over recent months between Bolshevik and Union Foresters, considerable optimism for cooperation and success still existed. Speaking just days before the law's issuance a member of the *Tsentral'naia Kollegia* (Central College) of *Narkomzem* made the following remarks:

> I know that many localities find the Basic Law On Forests not to their taste . . . . [B]ut in this case I will speak in the words of comrade Larin [it is unclear if this is the same Larin—Iurii—noted earlier in this chapter], 'If the localities do not accept the law they will have to be won over.' And I am certain that our comrade forest workers will win for themselves both the localities and the law itself.[103]

---

[101] "Vtoroi delegatskii s"ezd Soiuza lesovodov," 340.

[102] Ibid.

[103] See "S"ezd lesnykh komissarov." Stenogram, *Lesa respubliki* 8-9 (1918): 526.

VI
Implementing the "Basic Law on Forests"

In retrospect, the total reorganization of forest affairs envisioned in "On Forests" could hardly have been contemplated or attempted at a less auspicious time. As if popular discontent and the incompleteness of Soviet control over the countryside were not sufficient obstacles, the regime was just now embarking on a struggle for its very existence against White and Allied forces in the first battles of a brutal Civil War (1918-1921). That any attempt at forest reform was made in this atmosphere seems remarkable, a testament to the fortitude and optimism of the law's framers and champions. Still other challenges, however, were already either emerging or soon to do so. The fuel crisis, in particular, was worsening, placing ever-growing stresses on forests and simultaneously making conservation more and more impractical in the near term. And as will be seen, important weaknesses were soon to become manifest within the new structure of forest legislation and organization itself.

Foresters were, of course, well aware of many of these difficulties and explicitly tempered expectations accordingly. The anonymous author of an article published in *Lesa respubliki* a few weeks after the decree's issuance, for some of the reasons just noted, expected implementation to be "very complex." He thus emphasized that the new legislation was—as titled—only a "basic" law, "in the sense that it [did] not provide definitive, exhaustive statements on every single question" but served instead to "outline only the general nature of the future legal order."[1] Suggesting that full implementation was in any case not envisioned in the immediate future, the article also called for the center to focus on creating "instructions" to aid realization of a prioritized subset of twenty-seven basic "obligations" covering only about forty of the law's 120 articles. These focused on overarching considerations basic to the establishment of a planned forest economy, including organizing administrative organs and procedures, establishing quotas and norms for procurement and distribution of forest materials, creating and deploying cadres of technical forces, carrying out general inventorying, budgeting and accounting, and the excision of a small quantity of

[1] "Raboty Tsentra," *Lesa respubliki* 8-9 (1918): 477.

local holdings from the overall envisioned state forest in accordance with the law (see chapter 5). Though several conservation-related clauses were included in this framework, priority was given not to actual conservation work—which was, in fact, not mentioned explicitly at all—but simply to working out "norms for forest protection."[2] The need for a "complete hunting code" based on article 31 of the law was also noted.[3] Additional forward-looking notes on implementing other parts of "On Forests" were published in November. These envisioned, for example, the creation of local councils of communist forest specialists (*soiuzy lesnykh spetsialistov-kommunistov*), to be organized under a central counterpart, and the convocation of a Congress thereof.[4]

Early responses from around the country indicated that even these kinds of skeleton programs were unlikely to be realized quickly. Continuing a trend already long in process, some local organs, offended by centralist ambition, immediately and explicitly rejected "On Forests" entirely, claiming and acting upon the right to manage their own forest affairs in accordance with local wishes and needs. Such, for example, was the case in Olonets province, against whose "capitalist" and "restorationist" Land Committee a writer for *Lesa respubliki* railed in November 1918. Particularly galling in this particular case appears to have been the "treacherous" decree of a local Land Section (*Zemel'nyi otdel*) "increasing the size of forests of local use at the expense of former private, communal, and other forests, and insisting on the transfer of all of this to [their own] exclusive control." Rhetoric aside, these activities, of course, reflected SR rather than strictly "capitalist" tendencies.[5] To be sure, the actions and proposals of the Olonets Land Committee were condemned at a congress of Land Sections held in Petrograd during October. But beyond censure, there was little that could actually be done at this point. Nor was this an isolated occurence. In general, "On Forests" seems to have been about as well received around the country as had been centrally-directed efforts a few months earlier to transfer local forest

---

[2] Based on articles 29, 30 83 and 90 of "On Forests."

[3] "Raboty Tsentra," 477-82. The 42 prioritized articles are: 9, 17, 20-26, 29-34, 37-38, 44-47, 51-52, 71, 73-74, 83, 90-91, 94, 101-103, 106-114. A hunting code, in many ways comparable to "On Forests" was issued in 1919 under the title "Basic Law On Hunting."

[4] "Lesovody-kommunisty," *Lesa respubliki* 14-15 (1918): 752.

[5] Al'fa, "Ne priznaiushchye kommunisma," *Lesa respubliki* 16 (1918): 863-65.

responsibilities and inventory from Land Committees to Forest Sections (see chapter 5).

Here and there, however, there were positive, if modest, developments. The Soviet of an *uezd*-level forest sub-section (*podotdel*) in Vitebsk, for example, quickly "decreed to create a special poster about the law" to be disseminated around the *uezd* in effort to encourage its adoption. A copy was sent to the central forest apparatus for approval.[6] And throughout the summer and fall communications arrived in Moscow from local Soviets in various parts of the country about specific areas of forest recently "inventoried" or, sometimes, as "established" as of local use in accordance with the law. This latter development was not necessarily all that encouraging, however, especially from a conservationist point of view—forests of local use were areas to be *excluded* from the decree's overall planning and conservation priorities. As already noted, the intention had been that these exclusions, as exceptions to overall priorities, would be done later rather than sooner and on a very small scale. Obviously, however, these kinds of decentralizing aspects of the law were more instantly appealing to many local organs than their centralizing counterparts. They were thus implemented relatively quickly. Beyond these kinds of fairly specific communications, there were a larger number of enthusiastic, though vaguer, statements emanating both from the center and certain localities about "great work" done and "promising developments" already underway in the forests and around the country. Overall, however, these seem exhortative as much as descriptive.[7]

Implementation of the new law necessitated further legislative efforts on the part of the central organs. "Ruling on Accounting and Inventorying Former Privately-Owned Forests" (*Pravila ucheta i opisaniia byvshykh chastnovladel'cheskykh lesov*), worked up during June, for example, attempted to systematize the "procedural" activities of local organs working on the transfer of forests from private to state ownership. In this connection work was also begun to map and inventory accurately Russia's forest resources.[8] And throughout 1917

---

[6] N. A. Kuznetsov, "Propaganda osnovnogo zakona o lesakh," *Lesa respubliki* 16 (1918): 870.

[7] In just one of several typically vague examples, the Viatsk Forest Subsection (*Viatskyi lesnoi podotdel*) was praised in *Lesa respubliki* in November for having done "very much work" within the framework of the law. See, Tertius, "Revoliutsiia i mesta," 855.

[8] N. I. Faleev, "Chto sdelano i chto predstoit' sdelat'," 468.

and much of 1918 decrees were issued disbursing funds (usually inadequate, see below) for the support and salaries of forest guards and their work of protecting, inventorying, and so on, all within the framework of "On Forests."[9] Overall, however, legislative activity connected to implementing "On Forests"—and no doubt the general interest of the central government in forest conservation and organization—seems to have begun seriously to wane after mid-summer 1918 as priorities were switched increasingly toward solving the growing fuel crisis, which in turn was becoming fundamental to fighting the Civil War.

The first anniversary of the Bolshevik Revolution in November 1918 provided the new forest apparatus an opportunity to evaluate a hectic year's work. (By this time, it had undergone another name change: this time from *Tsentral'noe upravlenie lesov respubliki* (TsULR) to *Tsentral'nyi lesnoi otdel*—the Central Forest Section[10] or *Tsentroles* for short and hereafter. As before, it continued frequently to be referred to generically as the "forest department" or "Forest

[9] In January 1918, for example, even before the issuance of "On Forests," the SNK assigned for the basic "support" of the forest guard an annual sum of almost 21.5 million rubles ("Ob assignovke 24,390,577 rub. kommissariatu zemledeliia po vedomstsvu lesnogo departamenta dopolnitel'ym kreditom na uvelichenie okladov zhalovaniia lesnoi strazhe," *Dekrety Sovetskoi vlasti* 1: 562-63). The sum noted in the title of this decree is greater than that given in the text since the former covers a fourteen-month period beginning in November 1917 The figure can be compared with the much larger sums given regularly to the economic organs involved in forest exploitation (cutting and transport). One-hundred-thousand rubles, for example, was assigned March 1918 by the SNK to the Fuel Section (*Otdel topliva*) of the VSNKh for the 1917-1918 operating year to "strengthen the development of forest cutting and [timber] procurement" ("Postanovlenie ob assignovanii 100 mln. rub. Otdely topliva VSNKh dlia usileniia lesozagotovok v 1917-1918 gg.," ibid., 2: 28-29). A further 86.7 million rubles was assigned to the same organ in early June to cover shortfalls incurred the previous six months ("Postanovlenie ob otpuske 186,671,823 r. 63 k. otdelu topliva VSNKh na zagatovku lesa v pervom polugodii 1918 goda," ibid., 2: 609-10). Even these sums were greatly surpassed soon after. For the first half of 1919 a sum of one billion rubles was advanced ("Postanovlenie ob otpuske 1 mlrd. rub. VSNKh dlia Glavnogo lesnogo komiteta avansom v schet smety 1-i poloviny 1919 goda na lesozagotovitel'nye operatsii," ibid., 4: 684).

[10] The retitling occurred at some point between June and November. In writing and debate, however, it was not always respected at the time; thus one finds references to the central forest organ by both its old and new names—and both capitalized and not—throughout the period after the name change.

Department." A Congress of Land Sections under *Narkomzem* (*S"ezd Zemel'nykh otdelov pri NKZ*) held that month pointed to disappointing levels of implementation of communist forestry in general and resolved to redouble efforts.[11] Around the same time, Faleev, writing as usual in *Lesa respubliki*, praised the passage of the decree, by now some five months past, as "the greatest pride" and a signal achievement of Communist foresters; but he also complained there had been little substantial progress made in its implementation. Fundamentally, he argued, the general population remained hostile to the new forest order. Even the bulk of authorized forest workers, he asserted, themselves "failed" to understand its essential meaning or significance, which he identified as the shift from private to state ownership of forests. The creation of a unified forest plan encompassing and unifying production, distribution and protection, he further noted, remained largely paper achievements. Forest reality in November 1918 could still be described in terms of "separatism" rather than of unity and consensus. And still, he noted, there was a marked tendency "toward the reestablishment of private forest property."[12]

Forest conservation was given more serious consideration in early 1919 by a Congress of representatives of *uezd*-level sub-sections (*uezdnye podotdely*). This decreed the need for urgent measures to protect the forests, to which end it called in particular for the organization of an armed and "properly paid" forest guard, and for improved efforts to satisfy popular demands for timber through "organized" cutting "in accordance with national conditions." This, it was hoped, would reduce the incentive to rural populations to carry out their own, unauthorized procurements from already-exhausted local sources. These ideas, of course, were fundamentals of "On Forests" itself. The congress also tried to firm up enforcement of conservation law all around by decreeing that offenders should be turned over to the "people's court." First time offenders, however, were to be given the opportunity to "voluntarily pay" within five days a fine equal to twice the raw-material value of the timber felled (or thrice if the materials had already been used and could not be recovered).

As with all such calls, these too seem not to have had tangible results.[13] Though the ferocity of unauthorized cutting was reckoned on several occasions, particularly during 1918 and 1919, to have abated somewhat, its overall level

[11] "Resoliutsii s"ezda," *Lesa respubliki* 17-19 (1918-1919), 907.

[12] N. I. Faleev, "Godovshchina revoliutsii," 835.

[13] *Lesa respubliki,* 20 (1919): 1051.

remained distressingly high and almost completely immune to efforts to control it. Such declines as were noted, moreover, seem as often as not to have been the result of simple forest exhaustion rather than of any pro-conservation crackdowns or a change of heart among local populations. Faleev noted in 1919, for example, that in Kaluga province the "pressure on the forests" was finally "beginning to ease," that there were "no longer any mass cuttings," and that the population was "beginning to relate more peacefully to the [forest-related] directives issued by the [state] . . . ." On the other hand, he added, damage the previous year had been overwhelming. Every single linden tree in the area, down to the very last individual, had been chopped down, for example. Much of what remained was mere "bushes and shrub."[14] That cutting had slowed was thus understandable.

Throughout 1918-1921 local reports filed with the Foresters' Union—and numerous other forest sources—describe scant change in general conditions, and certainly nothing at all to cheer conservationists. Reports read at the Union's June 1920 Congress, for example, sounded almost identical to those of two years earlier. "Intense cutting" by "local populations" for domestic uses or for other purposes continued. So did the practice of sending livestock to graze on recently cleared areas in violation of law.[15] Woodlands around villages continued to thin, and in many cases had already become damaged beyond hope of near-term natural recovery.

What *was* news, perhaps, was the fact that *authorized* cutting, far from becoming increasingly sustainable, planned, or rational, was instead growing more destructive. In this case, forests along rivers and railroads were the main victims.[16] Here too, the anguish of foresters was almost tangible. "[E]verywhere in the Republic," it was asserted at the same Congress, unplanned and "uneconomical" [*bezkhoziastvennyi*] forest cutting" by state organs had become rampant.[17] Reforestation work, at the same time, was apparently going "poorly" (when it was going at all). Fires and insect infestations, facilitated by reckless

[14] Ibid., 1066.

[15] GARF. Fond 5520, del. 4, op. 1. st. 30. Depending on the specific area, this violated not only "On Forests" but all previous forest codes back at least as far as 1888. On the other hand, in response to peasant discontent, the state itself began gradually to roll back restrictions on grazing and other damaging activities beginning in 1919 (see elsewhere in this chapter and throughout chapter 7).

[16] Ibid.

[17] Ibid.

authorized cutting and a failure to do proper clear-ups, were rising, too.[18] The combined effect of state and popular depredations was summed up in stark and gloomy terms:

> The forest economy is completely ruined. . . . More accurately, there is taking place a genuine clearing [*svodka*] of the forest. Cutting sites [*lesoseki*] are not being cleared of debris and serve as breeding grounds for the spread of [harmful] insects. . . . Forests are failing to regenerate naturally due to the unauthorized grazing of cattle, the inadmissible breadth of cuts, the intensity of littering, and so on. The old stands are dying, new ones are not being produced. In a word, the forests are dying.[19]

Similar reports continued to arrive at the Union's central office throughout the year and beyond. Only the color of the language offered much variety.

The Fuel Crisis

Beyond peasant activities, which have already been discussed in some detail, several other factors were at work behind these developments, especially where the activities of state organs were concerned. In particular the country had run afoul of a crippling timber shortage or "fuel crisis." The impact on early Soviet forest use of the fuel crisis (or "fuel famine" as it was sometimes called) was enormous. Only its relatively short duration—essentially 1918-1919 (and to 1921 in attenuated form)—puts it behind peasant resistance as the chief overall obstacle to rational management and conservation. Two main consequences stand out: (1) greatly increased stress on easily accessible forests due to a massive drive by the state to cut and stockpile timber for fuel and building materials (an operation usually referred to as *zagotovka* or "stocking up"); and (2), as the supply of wood fuel became an ever greater national priority, the intrusion into forest affairs of other commissariats and a truly bewildering number of smaller organs in competition with *Tsentroles* of *Narkomzem*. Viewing the situation from the vantage point of 1923 one Soviet commentator put it this way:

---

[18] Ibid. During 1919-1921, 2 million *desiatinas* (about 2.2 million hectares) of Russian forest were damaged by fire. Over 1923-1925 the figure ran at about 1.4 million hectares a year. While these were high figures, they were far from unprecedented. In 1915, for example, a series of especially large blazes had harmed some 15 million hectares in Central Siberia.

[19] Ibid., st. 39.

> The [fuel crisis] . . . demanded heroic measures for the procurement of materials [and] created special timber-procurement organs which cut forest stands as close as possible to railroads and waterways. The Forest Department was forced to violate all forest-economic plans, to exhaust forest supplies and struggle with forest fires [exacerbated by ill-planned cutting].[20]

Both these changes obviously did huge harm to the cause of conservation: the first directly, both by increasing the amount of cutting and decreasing the care with which it was done; the second more indirectly by creating or facilitating various inter-departmental conflicts, tensions, and inefficiencies, tending to stifle protection initiatives and increase the lawlessness of operations in general. Chief among the "carpetbagger" organs in this respect was the main economic organ, the VSNKh—or more specifically its various timber supply subordinates, particularly the Fuel Section (*Toplivnoi otdel* or *Otdel topliva*) and, after December 1918, the Main Forest Section (*Glavnyi lesnoi komitet*), hereafter referred to as *Glavleskom*.

The Drive for *Zagotovka*

It has already been noted that during the first years following the Revolution, in consequence of territorial losses, damaged infrastructure, foreign embargoes, and other causes, Russia's dependence on wood-fuel, already very high, became overwhelming.[21] Fuel shortages hit many users hard, but the railroads in particular could not be allowed in a time of war to languish for lack of fuel. This meant running them almost exclusively on wood-fuel, which had obvious negative implications for any hopes for rational forestry or forest protection. Though of much less overall importance, wood was also used as an "indirect" fuel, with industrially useful combustible oils sometimes being

---

[20] "Lesnoi kodeks," *Lesnaia kooperatsiia* 10-11 (1923): 9.

[21] The impact of the Brest-Litovsk Treaty aside (see chapter 5), the Civil War also cut off or significantly reduced available supplies of coal from the Donets Basin for a protracted period, especially during 1919. Debate on this matter is preserved in RSFSR, *Tsentral'nyi Ispolnitel'nyi Komitet. I-IV sessii vserossiiskogo tsentral'nogo ispolnitel'nogo* komiteta. VIII sozyva. Stenograficheskii otchet (Moscow: Gosudarstvennoe izdatel'stvo, 1922), II sessiia, 52, and passim.

extracted by a process of partial burning.[22]

Though matters became critical only in 1918 and 1919, the drive for *zagotovka* in fact began in 1916 with the creation of new economic organs dedicated to war-time timber procurement and distribution. As noted, these had represented the first real challenges to what had been the more or less undisputed authority over all forest matters of the old Forest Department since its creation in 1798. The Bolsheviks built on this basis, creating further such organs from late 1917 on, and then giving them increasing power to cut and stockpile timber. Over time, these organs were brought under more and more tightly centralized authority. Among these early Bolshevik organs were the Timber and Peat Extraction Service (*Sluzhby lesnykh i torfianykh razrabotok*) and the Forest Supply Sections (*Otdely lesnykh zatogotovok*). The latter—sub-departments located within various local organs in a fashion similar to the deployment of Forest Sections within Land Committees and Soviets—were intended mainly to serve the needs of the railroads.[23]

In May 1918, at about the time "On Forests" debuted, these and other similar timber supply organs—including some left over from the period of the First World War—were in turn replaced or merged with special organs dedicated specifically to railroad fuel supply, which was becoming by far the chief concern. Entitled Railroad Timber-Supply Committees (*Zheleznodorozhnye lesozagotovitel'nye komitety*, or for short—and hereafter—*zheleskomy*), these were to have a considerably longer existence than their predecessors, and a major effect on many forest areas. Significantly, *zheleskomy* were subordinate neither to *Tsentroles* (within *Narkomzem*), nor to *Glavleskom* or its parent, the VSNKh (which as detailed further below, was by this time developing important jurisdiction of its own over supply-related forest affairs).[24] Officially, *zheleskomy*

---

[22] *V pomoshch' zhigariu: Pechnoe uglezhzhenie*. Prilozhenie k zhurnaly. *Les* 2 (Feb. 1917). Also published separately under same title at (Ekaterinburg: Izdanie zhurnalu "Les," 1917).

[23] RSFSR. Glavnyi Lesnoi Komitet. Upravlenie lesozagotovok. *God raboty po snabzheniiu zheleznykh dorog toplivom* (Gosudarstvennoe izdatel'stvo, 1920), 3.

[24] That these organs were independent of TsULR/*Tsentroles* (as the chief forest organ within *Narkomzem*, the Commissariat of agriculture, was variously named) is not surprising. More notable is their independence from *Glavleskom*, which was by definition the "top forest-supplies organ in the Republic" (ibid.,. ix, 3). In 1920, following protracted friction between *Glavleskom* and the *Zheleskomy* the latter were absorbed by the former. The source here cited, incidentally, provides much more detailed information on

functioned within a third People's Commissariat, that of Transport (*Narodnyi komissariat putei soobshcheniia*, or NKPS). In fact, however, they enjoyed considerable autonomy and—reflecting the growing bureaucratic complexity of forestry overall—had their own central organ, the Central Administration for Fuel and Forest Supplies (*Tsentral'noe upravlenie toplivom i lesozagotovkami*, or *Tsutop*).

Details aside, the overall pattern presented in the foregoing summary history of the organization of timber supply is one of increasing authority and autonomy for the organs charged with actually getting wood to the railroads—a clear reflection of the high priority accorded to timber supply and to a desire to effect this quickly and efficiently by reducing the bureaucratic constraints thereupon. Each step also represented further challenges to the forest authority of the official chief forest organ, Tsentroles, whose mandate to establish and administer forestry overall, including their conservation, became increasingly circumscribed. Thus, as the fuel crisis worsened throughout 1918 and 1919, numerous decrees and "orders" (*predpisaniia*) were issued by the SNK granting unfettered forest access to railroad and other users. One of 28 November, 1918, for example, ordered "forest sub-sections and land sections" (*lesnye podotdely i zemel'nye otdely*) to "carry out unrestrictedly and without compensation the release of forest materials to the Committee of State Construction and Public Works" (*Komitet gosudarstvennykh sooruzhenii i obshchestvennykh rabot*) for "strategic railroad works."[25] Another, in May 1919, required *Tsentroles* to "give over immediately to the use of *Glavleskom* [within the VSNKh] . . . 250,000 *desiatina*s of forest" with an "approximate supply of 6 million cubic sazhens situated in a ten-*verst* strip along both sides of [unspecified] railroad lines . . . ."[26] In July, *Tsentroles* was ordered to make another transfer to *Glavleskom*, this time of 375,000 *desiatina*s.[27] In the second instance, the target areas were all located along river banks. Forests along rivers and railroad lines were each targeted

---

these organs, and on railroad timber use in general, than has been reproduced here.

[25] "Predpisanie SNK lesnym podotdelam i zemel'nym otdelam gubispolkomov proizvodit' besprepiatsvenno i bezvozmezdno otpusk lesnykh materialov Komitetu gosudarstvennykh sooruzhenii i obshchestvennykh rabot dlia proizvodstva dorozhnkyh strategicheskykh rabot," *Dekrety Sovetskoi vlasti*, 4: 91-92.

[26] "Dekret SNK o poriadke otpuska lesa," ibid., 5: 147.

[27] "Dekret SNK ob opuske Glavnomu lesnomu komitetu 375 tys. desiatin lesosek pod zagotovky lesa," ibid., 338-39.

primarily because they were accessible to cut and easy to transport. Those located specifically along rivers, however, were at the same time exactly the kinds of erosion-sensitive areas frequently protected as "water protecting" (*vodookhrannye*) areas. (Though it is likely that forests officially thus designated were among those signed over for cutting by these and other orders, this cannot be verified here.) If these kinds of transfers seem compatible more with all-out exploitation than with planned or sustainable cutting, it should be noted that the last-cited decree did at least specify that *Tsentroles* should instruct its local organs to show *Glavleskom* where to cut. On the other hand, the latter was also assured permission to use still further areas if the need arose (article 2). Further such transfers were to be made over the coming year.

From 1918 to 1921 innumerable other decrees and measures were also resorted to in the search for more timber and better distribution. The conscription of individuals into the Red Army was foregone or postponed on several occasions in order to facilitate continued cutting, preparing, and transporting of timber, for example. Other decrees drafted persons directly into forest work itself.[28] At one particularly desperate moment in early 1919, the entire personnel of *Tsentroles* was drafted into the work of *zagotovka* (though they were required somehow to continue meeting their usual responsibilities also). A great many emergency quotas were issued for supplying certain major towns or areas with fuel. Special organs came and went with bewildering speed and variety. In June 1920, the peasant population around Moscow and nearby areas was drafted as a whole to deliver an urgently-needed 17,000 "kuby"[29] of timber to local organs. The following month, "shock brigades" (*udarnye gruppy*) were ordered created along

---

[28] See, for example: "Predpisanie Soveta Oborony vsem komissaram voennykh okrugov priostanovit' prizyv v Krasnuiu Armiiu i rezervnye chasti lits, zaniatykh na rabotakh po zagotovke topliva," ibid., 4: 143-44; and "Predpisanie SNK ispolnitel'nym komitetam Sovetov i sovnarkhozam Viatskoi, Kostromskoi, Severo-Dvinskoi, Vologodskoi, Cherepovetskoi, Olonetskoi, Petrogradskoi, Novgorodskoi, Pskovskoi, Tverskoi i Iaroslavskoi gubernii i komissaram zheleznykh dorog o vvedenii s 1 marta vseobshchei trudovoi povinnosti po zagotovke i dostavke drov dlia zheleznykh dorog," ibid., 630-32.

[29] It is unclear what the intended unit of measurement was here. The Russian term given—*kuby*—meant at different times cubic meters or cubic *sazhens* (an old Russian unit of measurement equal to about 28 inches). In other decrees around the same time units are sometimes defined specifically as cubic meters or cubic *sazhens*. The decree referred to here is "Postanovlenie STO ob ispol'zovanii letnogo perioda dlia snabzheniia drevesnym toplivom g. Moskvy," ibid., 9: 156-58.

key railroads to speed up the work of cutting and supplying timber.[30] War prisoners were put to similar use under this and comparable initiatives. In September 1920 and subsequently, school children were called on to collect pine cones for fuel.[31] And so on.

In response to these initiatives, actual cutting appears to have increased during the years of the fuel crisis at a fairly steady clip. During the 1919-1920 operating year, for example, approximately 700,000 "kuby" more timber was officially cut by authorized organs than during previous operating year.[32] This and other similar increases in subsequent years, moreover, were not the result of increasingly extensive timber harvesting practices—of the exploitation, say, of relatively untouched regions in the far north and Siberia—but rather of *intensified* cutting in traditionally-exploited regions in European Russia within easy reach of railroads, rivers, and villages. It was noted in 1923 in the All-Russian Soviet Central Executive Committee, for example, that between 1918 and 1923 some 243,545 *desiatina*s of forest had been thus cleared in European Russia, and subsequently turned over to agricultural use.[33] This figure, of course, covers only authorized clear-cutting—which may or may not include authorized "selective" cutting and certainly does not cover *any* type of cutting by *un*authorized parties. Figures released elsewhere in 1923 darkened the picture still further by highlighting the rampant—and apparently still unchecked—practice by state organs of "felling forest areas several decades ahead," effectively borrowing from projected future forest growth for immediate needs.[34]

Though official accounts record steady improvements in meeting railroads' demands for timber with each organizational reshuffle,[35] fuel shortages

[30] "Postanovlenie STO ob obrazovanii udarnykh grupp lesnykh komitetov," ibid., 202-04.

[31] "Postanovlenie SNK o sbore shishechnogo topliva v osennii period 1920 goda," ibid., 10: 164.

[32] RSFSR. *Tsentral'nyi Ispolnitel'nyi Komitet. I-IV sessii vserossiiskogo tsentral'nogo ispolnitel'nogo komiteta* VIII sozyva. Stenograficheskii otchet (Gosudarstvennoe izdatel'stvo, 1922), II sessiia, 30.

[33] RSFSR, *Vtoraia sessiia vserossiiskogo tsentral'nogo ispolnitel'nogo komiteta.* X sozyva. Zasedanie vtoroe (30 June). Stenograficheskii otchet (Gosudarstvennoe izdatel'stvo, 1923), 26.

[34] P. Serebrennikov, "O novom lesnom kodekse," *Lesopromyshlennoe delo* 17-18 (1923), 9.

[35] See, for example, RSFSR. Glavnyi Lesnoi Komitet . . . . *God raboty po snabzheniiu zheleznykh dorog toplivom*, ix, 3.

throughout the economy overall remained critical. During 1920, for example, available fuel (of all types) was estimated at only 20% - 50% of what was needed depending on the industry or sector involved. Overall, the railroads appear to have been given the lion's share of what was available.[36] Over the next couple of years shortfalls appear to have been trimmed to "only" 15-20%.[37] This was done largely by further increases in authorized cutting, again accounted for as "loans" against future years. *Glavleskom*, for example, as the chief procurement organ under the VSNKh, increased significantly the amount of timber it supplied to various users each year from 1918 to 1921. During the 1918-1919 operating year it "procured" 4.2 million "kuby"; in 1919-1920 the figure was 9.4 million; and in 1920-1921, 14.5 million. In each case, however, this was still below target cutting figures, which had been set at 9.4 million for the first year and 11.2 million for the second (no figure is available for the third year).[38] Almost all this timber appears to have come from forests in European Russia.

From a conservationist point of view it surely seemed as if the state, which had become the vessel for so many hopes for better use and protection, was instead becoming a major force for destruction. That this was more of necessity than of will, was, of course, clearly understood, and in many cases, accepted. A Congress of *Narkomzem*'s Land Sections (*S"ezd zemel'nykh otdelov pri NKZ*), held early in the fuel crisis (in November 1918)[39] noted for example:

---

[36] RSFSR. *Tsentral'nyi Ispolnitel'nyi Komitet. I-IV sessii vserossiiskogo tsentral'nogo ispolnitel'nogo* komiteta. VIII sozyva. Stenograficheskii otchet (Gosudarstvennoe izdatel'stvo, 1922). II sessiia, 25, passim.

[37] Ibid., 50.

[38] Ibid. Absolute shortages of coal, oil, timber, and turf were made worse by rampant thievery of existing supplies during transport and storage, as well as by wastage, such as timber left to rot after harvesting (ibid., 28, passim). Charges of theft were raised and denied on several occasions There seems, in fact, to have been some debate about the relevant importance of these various factors. In March 1920, having heard evidence relevant to each, the VTsIK, resolved that the "basic cause" of the fuel crisis was not bureaucratic problems or theft, but an "insufficiency of fuel, above all of coal," which it in turn traced to the "inadequate development of the planned economy" -- reinforcing thereby, the priority of carrying through the program of centralized planning envisioned in "On Forests" (ibid., "Postanovleniia sessiia [Prilozhenie], XIII). On "kuby," see n. 29.

[39]The Congress's main task relative to the forest economy concerned furthering the development of

> In Revolutionary times one cannot reckon with theoretical questions. Hardening our hearts, we must cut down our native, mighty forests, turning them into fuel . . . . [W]e must do this on behalf of that future which will grant us the possibility to busy ourselves only with our own direct work, to [properly] establish the forests, to rationally satisfy all the needs of the country, to create a communist basis for the forest economy, to cultivate timber and turn it into chemical and mechanical products of the highest economic value.[40]

Forest priorities were crystallized in similar language—but on a higher level—in a terse statement made in *Lesa respubliki* in early 1919: "The country is suffering through a fuel famine. Allocation of enormous quantities of forest areas for timber supplies and the realization of these supplies is demanded."[41] A 1922 essay in another forest journal, *Lesopromyshlennoe delo*[42], asserted that in the face of these kinds of priorities the "idea of rational use of timber" had in general been "cruelly shattered" in recent years, a development it explicitly blamed on the needs of the railroads in the context of the fuel crisis.[43]

Yet if conservation seemed in abeyance during the fuel crisis, it was clearly not entirely forgotten, not, at least, as a general principle and ultimate goal. This is the case regarding the above-noted Congress of Land Sections, which balanced its call to "cut down our native, mighty forests" with a demand that all Land Sections and Forest Sub-Sections "widely propagandize among the population" the "necessity of a careful attitude toward the forest," especially "protected"

Forest Sections within their parent Land Sections, and the latter's integration into the emerging general governmental structure. At this time, and typically, Forest Sections (*Lesnye otdely*) were located within *Sovdepy*, while Forest Sub-Sections (*Lesnye podotdely*) were deployed within Land Committees. Instructions written up towards this end by *Tsentroles* (of *Narkomzem*) were to be read here. All of this was in effort to realize article 6 of "On Forests," on the establishment of forest "technical organs." Further meetings of the Land Committees were held throughout the rest of the autumn (to December). Summary proceedings were published in *Lesa respubliki* in December 1918.

[40] "Lesnaia sektsiia: itogi rabot lesnoi sektsii," 906.

[41] "Uchet i mobilizatsiia lesnykh spetsialistov," *Lesa respubliki* 21-22 (1919): 1109.

[42] Published by *Glavleskom*, VSNKh.

[43] G. Kabanov, "Itogi S"ezda i zadachi Soiuza," *Lesopromyshlennoe delo* (no number, June-July 1922), 1.

forests.[44] It is the case also with numerous other congresses and conferences. And to some degree, it is true also of the state itself, which continued, at least on paper, to promote conservation. Instructions issued over Lenin's signature on 26 September, 1919, for example, required that efforts to supply the Moscow area with timber must not involve any cutting in "parks . . . culturally significant and protected areas" or in certain other off-limit places (article 1).[45] The instructions also specified an order of cutting based on the type and age of trees (article 3) and noted minimum densities of trees to be left after cutting in order to allow regeneration. Whether these instructions were actually observed, of course, cannot be determined here. There were other legislative efforts toward protection of this sort, too. A decree of May 3, 1920, "On the Protection of Green Spaces (Gardens, Parks, Suburban Forests and other Green areas)" called for inventorying such areas and establishing rules for their protection, primarily as areas of "exceptionally great significance for the health of the working population."[46] A "Decree on the Struggle against Forests Fires" was issued 28 July, 1920.[47] It confirmed the primary responsibility in this area of *Tsentroles*, called for the assignation of fire-fighting and prevention responsibilities in specific forests to particular local populations, urged better oversight, and so on. A more exhortative decree to "all provincial executive committees and communist organs" was issued in August, urging every effort to promote among the population a sense of concern and vigilance about fires.[48] Testifying to a sense

[44] "Resoliutsii s"ezda," 907. On the other hand, as the Congress continued into December, utilitarian priorities, not conservation, dominated the agenda. During this month, for example, it was decided that henceforth there was no place for forest preservation on the meeting's agenda at all ("Vserossiiskii s"ezd zemel'nykh otdelov, komitetov bednoty i kommun. Protokoly," *Lesa respubliki* 17-19 (1918-19), 942.

[45] "Instruktsiia SNK k dekretu SNK ot 14 avgusta o vyrubke lesov, nakhodiashchikhsia v 30-verstnoi polose Moskovkoi okruzhnoi zheleznoi dorogi," *Dekrety Sovetskoi vlasti*, 6: 139-40.

[46] "Ob okhrane zelenoi ploshchadi (sadov, parkov, prigorodnykh lesov i drugikh zelenykh nasazhdenii)," 29 April, 1920. *Sobranie uzakonenii i rasporiazhenii raboche-krestianskogo pravitel'stva* (Moscow: Gosudarstvennoe izdatel'stvo, 1920), 32: 132. This was followed on 16 September, 1921 by a decree "Ob okhrane pamiatnikov prirody, sadov i parkov." Both these decrees are relevant more to the story of zapovednik-style conservation than to the account of forestry given here.

[47] "Postanovlenie STO o bor'be s lesnymi pozharami," *Dekrety Sovetskoi* vlasti, 9: 280-82.

[48] "Postanovlenie Presidiuma VTsIK i Tsentral'nogo Komiteta RKP(b) o bor'be s pozharami,"

of crisis, provoked by an unusual number of fires (itself the result of a complex of factors including dry weather, poor cleanup after cutting, complacence, etc.), a companion decree addressed to "all citizens of the RSFSR" was issued the same day.[49] Only as the fuel crisis began to abate—and especially once the Civil War had been won—however, did real signs begin to emerge of a shift in state cutting priorities (see chapter 7).

Bureaucratic and Jurisdictional Issues

As noted, the realities of the fuel crisis, and the great priority put by the Soviet government on increasing wood supplies, stimulated rapid and important changes in the organizational structure of the forest regime only recently set up in "On Forests." These changes, in turn, were to have important consequences for the implementation of the decree in general, and of its conservation measures in particular. The general trend in this respect was the continuing demotion of *Tsentroles* and the elevation of the VSNKh (*Glavleskom*'s parent). This, in turn, led to increased bureaucratic overlaps, tensions, and inefficiencies.

The key relegation of *Tsentroles* in this regard appears to have resulted from the "Decree Regarding Forest Procurements for the Needs of the Railroads." This was passed over Lenin's signature on 2 August, 1918.[50] Issued "in light of the critical fuel situation on the railroads," it obliged three organs to take on "independent fulfilment of wood fuel procurement on the very broadest scale for the needs of the railroads." These organs were *Tsentroles*, the VSNKh (at this time in the form of its subordinate procurement organ, the Fuel Section [*Otdel topliva*]), and the Commissariat of Transport. All Soviets, Forest Sections, and Land Sections were called on to offer "full cooperation." Significantly, however, among the three specified organs, the VSNKh was made a kind of *primus inter pares* with the right of general oversight of the others (article 2).

Tsentroles's role in procurement-related affairs declined further only a few weeks later, with the creation of the "troika."[51] This three-man commission,

---

Dekrety Sovetskoi vlasti, 10: 41-42.

[49] "Vozzvanie VTsIK 'ko vsem grazhdanam RSFSR' o bor'be s pozharami. Ibid., 42-43.

[50] "Dekret o lesnkykh zagotovok dlia nuzhd zheleznykh dorog," *Dekrety Sovetskoi* vlasti, 3: 133-34.

[51] "Postanovlenie o poriadke finansirovaniia na zagotovky drov i o sozdanii komissii dlia zagotovki drevesnogo topliva." Ibid., 254-55.

comprising one representative each from the Commissariats of Transport (the *Narodnyi kommissariat putei soobshcheniia*), Agriculture (*Narkomzem*), and the Fuel Section of the VSNKh, was intended to consolidate the activities of all the above organs' fuel supply-related activities. To this end it was granted "dictatorial power" over all procurement affairs, effectively removing whatever authority *Tsentroles* still had in this arena.[52] The *troika* was required by the SNK to "prepare and transport as quickly as possible the greatest quantity of wood fuel, bring into this work all organizations and establishments which can assist in this work." "All Soviet organizations" were required under the threat of "strict responsibility" to render all possible assistance. The *troika* was empowered to "requisition all wood supplies, machinery, and tools necessary for wood fuel procurement" (article 2). All subsequent funding for these activities appears to have passed through it over the next several months.

Tsentroles's expulsion from procurement matters appears to have been completed by the issuance on December 1918 of a "Decree of the SNK on the establishment of the Main Forest Committee [*Glavnyi lesnoi komitet,* the aforementioned *Glavleskom*]."[53] Lenin was actively involved in the creation of the decree and was its chief signatory. Based on article 72 of "On Forests" (which had deferred to a separate decree the exact structure and composition of the highest levels of central forest authority), it established *Glavleskom* within the VSNKh and placed under its command "all central establishments which have managed or are managing . . . the forest economy" (article B2). These included the above-noted *troika*, which appears to have ceased any separate existence at this point. *Tsentroles*, along with its local organs at all levels, while not extinguished by this decree, was ordered "in all affairs concerning procurement and forest transport" to "completely subordinate itself" to the directives of *Glavleskom* (article B3)." Forest activities involving cultivation, simple administration, taxation, education—and all types of technical work—remained with *Tsentroles* and its parent, *Narkomzem* (article 3b). *Glavleskom* itself was to consist of six persons, five to be appointed by the VSNKh and just one by *Narkomzem* (article 5b).

---

[52] The original members of this commission, were: Volkovskii (Transport), Lanin (Fuel Section, Agriculture), and Golub' (*VSNKh*).

[53] "Postonavlenie Soveta narodnykh kommissarov ob uchrezhdenii Glavnogo lesnogo komiteta," *Dekrety Sovetskoi* vlasti, 4: 262. Two versions of the decree exist, one by the *Sovet oborony*, the other by the *Sovet narodnykh komissarov*. The latter appears to be the one actually decreed.

The wresting of timber procurement out of the hands of *Tsentroles* (and thus of *Narkomzem*) was in fact only part of a larger, ongoing reorganization of forestry which involved essentially splitting it entirely into two separate realms. These had already come by now to be referred to as *lesokhoziastvo* and *lesopromyshlennost'*—literally "forest economy" and "forest industry," respectively. The former, covering general planning and administration, as well as conservation and reforestation, remained through the 1920s and beyond under the jurisdiction of the *Tsentroles* (and its successor organs within *Narkomzem*).[54] The latter—focused currently on wood fuel procurement, but including also all categories of industrial supply—became ever more firmly established under the jurisdiction of the VSNKh and its subordinate organs. If this seems at first glance an equal split, it was far from such—specifically because during the fuel crisis in particular, but under Soviet authority in general, *lesopromyshlennost'*—the industrial use of forests—retained a much higher relative importance.

In November 1919 the need to increase timber supplies became so acute that, as already noted, all forest personnel and organs under *Narkomzem*, including *Tsentroles*, of course, were drafted into this work.[55] In fact, this draft seems to have been difficult to enforce, provoking further circulars and at least one decree angrily calling for *lesnichye* in particular to take the whole affair more seriously.[56] Nonetheless, *Glavleskom* continued to accrue greater and greater authority within its realm of timber procurement and supply. During 1918-1919, the entire annual forest cutting area [*lesosek*] was officially placed exclusively in its hands. (In fact, however, and insofar as *Glavleskom* lacked sufficient resources

[54] Though literally meaning "forest economy," given the activities it involved, "lesokhoziaistvo" is better—and henceforth translated as "forest planning." *Lesokhoziaistvo* is defined elsewhere, for example, as involving "inventorying [and] protecting [forests], taking care of planted areas, raising the productivity of forests, and organizing their manifold uses." See *Lesnoe khoziaistvo SSSR za 50 let, 1917-1967* (Moscow: Lesnaia promyshlennost', 1967, 11.

[55] "Dekret SNK o privlechenii k drovianym zagotovkam vsekh lesnkykh organov Narodnogo komissariata zemledeliia na osnovanii polnoi militarizatsii sotrudnikov lesnogo vedomstva," *Dekrety Sovetskoi* vlasti, 6: 305-07. A second decree issued a few days later (29 November) further clarified the nature of this draft, explicitly calling up, "regardless of age," all personnel from *lesnichye* on down ("Instruktsiia SNK o drovianykh zogotovok," Ibid., 321-24).

[56] See, in the latter case, "Postanovlenie Soveta Oborony o privlechenii lesnichykh k lesozagotovitel'nym rabotam," *Dekrety Sovetskoi* vlasti, 7: 551-54.

fully to make use of this power, some forest areas were given over to *Tsentroles*, perpetuating tensions between the two organs.[57])

The case of *Tsentroles* and *Glavleskom* was the most obvious forest-related bureaucratic overlap, but not the only one. The forest role of the Commissariat of Transport, for example, has already been noted. And under the conditions of the fuel crisis even the Commissariat of Internal Affairs (*Narodnyi kommissariat vnutrennykh del*) became a forest organ—in May 1919 the Soviet of Worker-Peasant Defense (*Sovet raboche-krestianskoi oborony*), decreed the creation nationwide of "temporary artels for fuel procurement and transport." A three-man "special commission" within the Commissariat of Internal Affairs was created to unify and direct these *artels*' energies.[58] Each of these developments offended some office or other, contributing to the general confusion.

In a variety of forums, the problems caused by these kinds of overlaps was noted and condemned. The All-Russian Congress of Land Sections, Committees of the Poor, and Communes (*Vserossiiskii s"ezd zemel'nykh otdelov, komitetov bednoty i kommun*), held in December 1918, for example, took up the issue with great interest,[59] seeing herein some of the real reasons implementation of "On Forests" had so far proved elusive. Its second resolution noted, for example:

> that the catastrophic condition of our forest economy, [while] connected with the general economic ruin engendered by many years of war, is to a significant degree the result of the indecisive policies of the leaders of the forest department, which, having allowed the formation of a whole series of parallel forest organizations, not connected with the peasant masses, and stuck in the bureaucratic methods of the departmental offices, has preferred the route of behind-the-scenes deals and interdepartmental friction to the healthy revolutionary policy of relying on the will of the organized proletariat[.] [This] has brought the entire edifice of forest politics towards total breakdown.[60]

While the faith placed here in ordinary Russians might in retrospect seem

---

[57] See GARF. Fond 5520, op. 3, del 1, st. 37.

[58] "Postanovlenie Soveta Oborony o merakh po zagotovke i dostavke topliva," *Dekrety Sovetskoi vlasti*, 5: 200-01.

[59] *Lesa respubliki* 17-19 (1919): 958-84.

[60] "Vserossiiskii s"ezd zemel'nykh otdelov, komitetov bednoty i kommun. Rezoliutsii," *Lesa respubliki* 17-19 (1919): 981.

somewhat misplaced, several other speakers at the same forum concurred, at least regarding the need for bureaucratic simplification. One condemned the current "coalition" of forest organs and called for a transfer of "the whole forest apparatus" to "a single pair of hands."[61] Faleev, also present, himself criticized the "uncoordinated activities of various departmental procurement organizations."[62] The congress also passed resolutions calling on *Narkomzem* and the Presidium of the VSNKh to resolve these issues, specifically by "work[ing] out a project for a decree regarding the creation of a single supreme forest organ for the entire country." The project, the resolution continued, should "unite all *lesokhoziaistvennye* . . . and *lesozagotovitel'nye* functions" (i.e. all forest-economic—especially forest planning—and all procurement functions), should unify interests and operations of rural and urban forest economies, and [should] satisfy all needs of the country's industries and enterprises for timber.[63]

Recognizing these problems, however, was easier than solving them, since this only led to further friction over whom to empower as the ultimate source of forest authority—or in the event a merger be recommended, how to effect this. Personnel within *Tsentroles* and its dependent organs cited the text of "On Forests"—which clearly favored their organization. From the economic organs under VSNKh came claims the former had already proven their ineptitude. "What does the Central Forest Section [*Tsentroles*] do?" asked one critic, Lavrov, in 1922,

> Forest-planting? Ask it how many *desiatina*s of land it has planted over the revolutionary period. Forest protection? How has this been manifest? Forest-inventorying work? Let it concretely point out what has been done and where. [*Tsentroles*] has not carried out, and cannot carry out any work [at all] in the broad sense of the word as long as it remains part of *Narkomzem*."[64]

Thus, while the solution to bureaucratic overlaps and inefficiencies was seen by virtually all interested parties as a merger of all forest organs and functions under one common organ—preferably at full Commissariat level, and thus independent of *Narkomzem*, the *VSNKh,* and the rest—much more

---

61 Ibid., 957-59. The speaker is one Vaniukov, from Kaluga province.

62 Ibid.

63 Ibid., 959-60.

64 I. Lavrov, "Mezhdu dvumia s"ezdami," *Biulleten' tsentral'nogo upravleniia lesnoi promyshlennosti (glavleskoma) i tsentral'nogo komiteta profsoiuza derevoobdelchikov* 10-11 (1922), 12.

problematic was the question *how* the existing organs of forestry ought to fit into this. Personnel within *Tsentroles* and those within *Glavleskom* each felt their own apparatus should dominate the merger and take over responsibilities previously assigned to the other. Lavrov, for example, championing the interests of *Glavleskom*, asserted that all forest tasks—planning, exploitation, and conservation—were seamlessly linked and should thus all come under *Glavleskom*'s control. Proponents of *Tsentroles* and *Narkomzem* used similar reasoning to stake their own claim to full control. [65]

In fact, forest-bureaucratic confusions and tensions were even more complicated than already indicated for at least two reasons. First, not only current overlaps, but a host of older jealousies poisoned relations between *Narkomzem* and the VSNKh. Much of this seems to have been connected with *Narkomzem*'s near monopoly of the creation of "On Forests" back in 1917-1918 (the subsequent decline of *Narkomzem*'s forest authority relative to that of the VSNKh seems only to have roused passions further). Through 1919 at least, various individuals within or connected to the VSNKh expressed their dissatisfaction with the almost total domination of that the process by Faleev in particular and the agricultural organ in general.[66] Second, even within Commissariats there appear to have been tensions, particularly among the various levels of sub-organ within the hierarchy under the VSNKh. Here, provincial-level organs of *Glavleskom* resented the authority of their parent, while both of these strained at the bit under the *VSNKh*. These and other tensions could not but hurt the efficiency of forestry in general, especially at a time when decisive action was vital if the reigning state of chaos in the forests was to be brought under control. The campaign to unify forest authority and administration continued in various forms through the 1920s and beyond, and cannot be traced here. Through at least the mid-1920s, however, the matter remained in the realm of talk, plans, and hopes. Ultimately, only in the Ukraine was a full Peoples' Commissariat of Forestry—*Narkomles*—to be created; and even this was short-lived and apparently of little real significance.[67]

The fuel crisis and related bureaucratic inefficiencies aside, other problems,

---

[65] Ibid., 13.

[66] Vserossiiskii s"ezd zemel'nykh otdelov, komitetov bednoty i kommun, Rezoliutsii," 962-64.

[67] V. E. Boreiko, "Plakala Sasha kak les vyrubili," 32.

too, plagued efforts and deflated hopes to implement "On Forests." There was, for example, a continual lack of properly-trained personnel at all levels in both the center and localities. Learned foresters in particular bitterly criticized recently-appointed replacements, assistants, or commissars as amateurs and bunglers, as "simply yesterday's *konduktor* (a low-level, relatively un-educated bureaucrat) elevated to the rank of forester [*lesovod*] at the will of the revolutionary movement," and as individuals largely or even entirely lacking in the necessary scientific training. A letter of 1920 sent to the Central Soviet of the Foresters' Union from its local branch office in Voronezh province, illustrates this particular tension perfectly. It began in all-too-familiar manner: the forests in the province, were "currently in the most mournful state"; in recent years, felling had reached "incredible proportions"; "unauthorized cutting and damage caused by cattle-grazing" had risen "to an unprecedented level"; and so on. The moment, the letter continued, clearly called for forest personnel with "indomitable energy, great organizational talent, strong will, persistent character, [and] great theoretical and practical preparation." These characteristics were in desperately short supply, however. As of the previous September (1919), Voronezh province had become a "front line" in the Civil War. Consequently, many of its best personnel had been either evacuated or gone into military service. Some, it seems, had joined the White armies. Coming on top of earlier losses incurred as a result of local hostilities to specialists in general, this had made matters intolerable. Though replacements had been provided, most were "young and inexperienced." Worse, the local forest organ[68] had by order of the local Soviet been put under the control of one "comrade Novoskol'tsev." In the opinion of the remaining members of the local Foresters' Union, this was a man

> totally incompetent in [matters of] forest economy, poorly educated, more or less backward [*nerazvityi*], with a temperamental character, erratic, [a person who] everywhere, regardless of place or company speaks in the most coarse and vulgar fashion, [who] speaks in the familiar to all his subordinates—even forest specialists [*lesnichye*] who are superior to him not only in age but also by their education and length of service. Comrade Novoskol'tsev . . . having declared himself the all-powerful boss of forest affairs, personally designates and transfers *lesnichye* and their comrades, makes his own selections to the office of area inspector, [and does so] not only without conferring with the local branch of [the Foresters' Union], but indeed deliberately and persistently ignores [such assistance] . . . .[69]

---

[68] The Voronezh Provincial Forest Sub-Section (*Voronezhskyi gublespodotdel*).

[69] GARF. Fond 5520, op. 4, del. 8, st. 1.

Novoskol'stev, it was claimed, had even begun to create something of a local dynasty, appointing numerous favorites—"unsuitable persons"—in "many [other local] *lesnichestva.*" The case of Novoskol'stev appears to have included a heavy element of class bitterness, pitting old elites against new recruits, and perhaps a more refined and urbane older generation against much rougher revolutionaries and peasants. But more than cultural conflict was at issue. That there was a real decline in the overall quality of forest specialists and training, at least for some years after the Revolution, seems unassailable. The sheer number of dismissals of experienced persons made this inevitable. Numerous contemporary assessments made an identical point. A writer for *Lesopromyshlennoe delo* asserted in 1922, for example, that the technical expertise of *Tsentroles* in particular had been "hugely weakened and damaged" since the Revolution.[70] And throughout the early years of Soviet rule, complaints poured in from specialists all over about numerous "Novoskol'tsevs" rapidly taking up influential forestry positions to the detriment of the quality of forest organs and forest work.

The Bolshevik authorities themselves appear to have viewed this process with some ambiguity. At the very top, as has already been made clear, Lenin understood the value of specialists and opposed efforts to terrorize and dislodge them, though he did also seek to subordinate them to political control. Lenin's circular of 4 April, 1918 defending specialists has already been noted in this regard. On 25 January, 1919 the SNK issued another decree (also over Lenin's signature) with the same aim. Entitled "On Registering and Mobilizing the Republic's Agricultural-Economy Specialists" *(Ob uchete i mobilizatsii spetsialistov sel'skogo khoziaistva respubliki)*, it allowed the state to draft into forest work anyone with certain basic relevant qualifications or experience, including, apparently, numbers of individuals who had recently quit forest work due in part to the unpleasant and thankless conditions of the job.[71]

Further down the Bolshevik chain of command, however, the value of specialists seems to have been less appreciated. *Lesa respubliki*, whose anti-specialist animus has already been noted, typically dismissed all complaints as

[70] K. B., "K voprosu o edinom lesnom organe," *Lesopromyshlennoe delo* 7-8 (1922): 5. As a journal connected with the VSNKh, which sometimes bumped up against *Narkomzem* on matters of forestry jurisdiction, *Lesopromyshlennoe delo* may have had its own reasons for criticizing the quality of the personnel in the agricultural and forest organs.

[71] "Uchet i mobilizatsiia lesnykh spetsialistov," 1109.

sour grapes and elitist resentment on the part of "old bureaucrats" and "bourgeois specialists." Their replacements, it argued, were "vibrant" individuals, representatives of a "green youth" which, even if it did "sometimes made mistakes," even "gross ones," did so "precisely because [it] create[d] and work[ed]."[72] The leadership of the Foresters' Union, in contrast, did nothing but make "complaints" and engage in "constant whining."[73]

Considerable further damage was done to the ranks of forest specialists by the personal decision of many highly-qualified individuals either to emigrate or even to go over to any of the various anti-Bolshevik forces in circulation at the time. An article published in *Lesa respubliki* during the autumn of 1918 named and condemned several "forest Whiteguards" [*lesnye belogvardye* (sic)]. A. T. Molchanov, a recent vice-director of the old forest department, and one "Burlakov"[74] had apparently fled to join the Ukrainian hetman P. P Skoropadskii as forest experts, for example. One Professor Ogievskii of the Petrograd Forest Institute also fled around this time to the Ukraine. Even G. F. Morozov, one of Russia's most celebrated foresters then or since, was also reported in 1918 as having ended a lengthy vacation in the Crimea to lend his expertise to the "Skoropadshchina"(as Skoropadskii's right-wing regime and activities were labelled).[75] Morozov even wrote and published a book entitled "How to Defeat the Bolsheviks" (*Kak pobedit' Bol'shevikov*).

Still further challenges were presented to conservation in particular by the perpetual lack of adequate funds. Several factors appear to have been at work here. For one, the termination of commercial intercourse with most of the outside world that followed the Bolshevik seizure of power hit the forest economy, which had previously been a major source of export revenues, particularly hard. Second, and as already discussed, during the years of the fuel crisis—though to a significant degree before and after this also—state priorities directed available resources not at conservation but at exploitation. Third, the state-wide policy of *khozrashchet*, or "self-accounting," introduced during the period of War

---

[72] Tertius, "Revoliutsia i mesta," 856-57.

[73] Ibid. See also *Lesa respubliki* 14-15 (1918): 743-44.

[74] A man by the name of Burlakov is noted in chapter 7 as having spoken on forest matters before the Soviet Central Executive Committee in 1924. It is unlikely this is the same man, though this cannot be established here.

[75] "Lesnaia belaia gvardiia," *Lesa respubliki* 14-15 (1918): 749-51.

Communism—also negatively impacted forestry in general and conservation in particular. *Khozrashchet*'s primary requirement, that all sectors of the economy be self-financing, presented particular problems. In brief, timber users lacked the funds to pay for forest products—and thereby could not provide forest organs' income. Yet the sector was indispensable throughout the economy, so could not be allowed simply to wither or die. Overall, it seems, forestry simply limped on, perpetually chronically underfunded.

Looking back in 1924, a participant in a forest debate at the Soviet Central Executive Committee (by the name of Kozyrev) outlined some of the chief consequences of underfunding on the forest economy over the past few years. (Even in the greatly improved budgetary conditions of the then current operating year, 1923-1924, forest funds were still only 25% of the pre-war amount.[76]) "During the years of the Revolution," he noted, and "over recent years, in connection with the general reduction of credit . . . work aimed at establishing [proper] forest management and forest technical work [has] ground to a complete halt." The "few traces of [positive] technical impact" established prior to 1917, on the forest economy had, he claimed, over the following few years been "completely swept away." "One can say that over the period 1917-1923 . . . land organs have not been able even to think about organizing a proper forest economy, but have been forced to think only about somehow protecting the forest economy from further damage."[77] The "profound upset to the forest economy" thus engendered, Kozyrev reckoned, would take "not just a year or two" of better funding to overcome.[78]

Among the most acute funding-related problems directly impacting efforts to implement "On Forests'" conservation initiatives were the salaries of forest workers. Prior to World War I the average monthly salary of a "qualified" forest worker was 127 rubles and 97 kopecks; that of a simple forest guard was 17 rubles and 17 kopecks. As late as 1924 *Narkomzem*'s budget allowed only 14 rubles for "qualified" workers and 5 for guards. Pay during the Civil War appears to have been about the same, though more intermittent. In 1924 Forest guards' pay was reckoned to be less than that of "workers and employees in all other

[76] RSFSR. *Vserossiiskii tsentral'nyi ispolnitel'nyi komitet.* XI sozyv. Tret'ia sessiia. Zasedanie vos'moe (15 October, 1924), 353.

[77] Ibid., 352.

[78] Ibid.

industrial sectors of the people's economy."[79] That said, local variations around these averages were great.[80]

A sense of what could have been done had funding—especially of the Forest Guard—been higher and more reliable is given by comments made by a local representative of *Tsentroles*, also in 1924. The man remarked at a meeting of the All-Russian Soviet Central Executive Committee that the previous year, "in the face of huge unauthorized cutting" in the Severo-dvinsk province, the local guard had begun "a resolute struggle to liquidate unauthorized cutting." Illegally-gotten materials were "confiscated or requisitioned," and so on.

> But these measures could not be [properly] followed through because of the lack of resources. We had to start on a course of illegal measures in order to carry out our task. We turned [for money] to *Narkomzem*, to *Narkomfin* [the Commissariat of Finance], but without result. Then by [local] decree of the *Ispolkom* we took 2,000 rubles from [monies allocated to] . . . forest clear-up [work]. Fortunately, the *Narkomfin* inspector never came round, otherwise he could have turned us in for these illegal actions. Over a period of five months we eradicated the problem, requisitioned the illegally-cut materials and returned to *Narkomfin* more than 20,000 rubles."[81]

It is to be assumed that the illegal requisitioning of funds for forest conservation by local guards—apparently so effective here—was far from typical. More often, the shortfalls simply meant less guarding and worse abuses. In many cases, underpaid forest guards spent much of their time not at their assigned tasks at all, but growing potatoes or other crops to feed themselves and their families. Others simply took bribes and looked away as illegal cutting went on.

Yet, despite all these problems, still some progress was being made during the very early 1920s in implementing aspects of "On Forests." If some local organs continued to stymie implementation, others worked in the opposite direction. A March 1920 decree of the Nizhegorodskii provincial executive committee, for example, cited articles 40 and 41 of "On Forests" as the basis for drafting "all [local] citizens . . . into broad participation in assisting the forest sections [*lesnye otdely*] and their organs [in] carrying out a merciless war against

---

[79] Ibid., 355.

[80] One speaker even claimed that forest guards were actually better off than most Russians, since although they got paid almost nothing officially they took huge amounts of bribes. Ibid., 369.

[81] Ibid., 372-73.

all plundering of the forest and all unauthorized felling.[82] In case of violations, the law went beyond its parent, and called not only for "full payment of the established fines," but also for "suspension of receipt of fuel and building materials for a period of one year." Growing increasingly draconian clause by clause, the decree even went on to state that in the event the guilty parties could not be found, the losses caused by illegal cutting would be "imposed upon all citizens of the village located nearest the site of the cutting . . . ." (That this might simply encourage people to plunder a little further from home seems not have been a concern.) Citing clauses 26 and 37, the decree carried on to draft all citizens, "of both sexes between the ages of sixteen and fifty-five" into work "clearing the forests of debris and raising [the level] of forest restoration." Draftees were also envisioned for "general labor duty [*trudovaia povinnost'*]," norms of which were to be set jointly by the region's foresters, the local *ispolkom*, and Soviets and Committees of Labor [*Sovety i komitety truda*] and which was to be carried out during two established seasons, each of six weeks' duration, in the spring and fall. For all of this work "no . . . special compensation" was to be provided, though workers would be entitled to a 50% discount when purchasing timber or wood fuel supplies within their *lesnichestvo*. Failure to fulfil one's turn at labor duty, if occasioned by "legitimate causes" resulted in a loss of the discount. Absence without "legitimate cause" resulted in loss of the right to receive any wood fuel or building timber at all for a period of one year.[83] This seems not to have been an isolated example.

At the same time, the Foresters' Union, whose political clout had suffered irreversible damage since the Revolution (due to a variety of factors including its explicit support for the Provisional Government prior to the October Revolution and its general status as an organization of technical rather than political experts), nonetheless also tried to remain actively engaged within the new order, and in particular to help reestablish a viable forest preservation regime following the recent waves of dismissals of forest guards. A "Decree on the Forest Guard" [*Polozhenie o lesnoi strazhi*], worked up by the union's central soviet and designed to operate within the parameters of "On Forests," provided for maintaining the forest guard institute [*Institut lesnoi strazhi*] which, as part of the state's ongoing effort to break up and subordinate older cadres of forest

---

[82] GARF. Fond 5520, op 4, del. 8, st. 5.

[83] Ibid.

personnel, was then under threat of extinction. The Union's plan, which seems to have envisioned preserving most of the existing personnel, as well as some level of status and authority for its leadership, proposed splitting the guard in two, with one half devoted to "technical matters" (and subordinated to political authority) and the other half given a freer hand to devote itself entirely to forest preservation. The plan seems not to have been all that successful, however. Though the specific twists and turns of the fate of the Forest Guard are beyond the boundaries of this study, it does appear more and more to have become exactly the kind of subordinate technical organ envisioned by Bolshevik authorities.[84] The general organizational form, duties, and status of the new Forest Guard were given fairly definitive shape by a combined decree of the VTsIK and SNK, in agreement with several other organs, in March 1923 (see chapter 7).

Other encouraging achievements in early implementation of "On Forests" can be recorded too. Regarding conservation, and solely by organs under *Narkomzem* (primarily *Tsentroles*), more than 15,000 *desiatina*s had been secured from immediate danger of erosion by forest preservation or reforestation measures. Complete replanting had been carried out on 18,000 *desiatina*s with an additional 58,000 partially replanted to increase the density of existing forest land. "Care of young shoots" had been carried out on 72,000 *desiatina*s. Prior to the Bolshevik Revolution, according to Soviet figures, the center had controlled some 100 million *desiatina*s of forest divided into 858 *lesnichestva* with 2,026 technical personnel. By 1920 this had increased to 140 million *desiatina*s divided into 2,485 *lesnichestva* with 6,500 technical personnel (the quality of these personnel, of course, is questionable).[85] By 1920 some 15 million *desiatina*s

---

[84] GARF. Fond 5520, op. 2, del. 21, st. 124-25. The exclusively preservationist wing of the forest guards would itself consist of two types of personnel: guards proper (directly responsible for preservation), and *ob'ezdchiki* (or "patrollers," charged with touring the forests regularly and reporting on their condition). Furthermore, all new hires, the draft asserted, must be literate, and each guard should be supplied with a firearm. (An order issued by the Revolutionary War Council [*Revoliutsionnyi Voennyi Sovet*] on 24 March, 1919 authorized the issuance and carriage of guns by forest workers. Little seems to have come of it however, judging by the passage of a resolution at the Foresters' Union's Third Delegates Congress in June 1919, demanding implementation of this order. See GARF. Fond 5520, op. 3, del. 1, st. 53. The draft also stressed the Foresters' Union's right to oversee the hiring and firing of staff, and the need to pay guards' salaries from state funds.)

[85] B. N. Knipovich, *Ocherk Deiatel'nosti NKZ-a,* 37.

(about 16 ½ million hectares) of formerly private forest had been taken over and 1 million thereof inventoried. Less encouragingly for conservation, though relevant nonetheless to implementation of "On Forests," also by 1920 some 4 million *desiatinas* (about 4.4 million hectares) had been assigned for cutting for state needs; and 30,000 had been approved to be clear-cut and turned to agricultural use.[86]

These, however, were modest achievements at best. By the end of the Civil War in 1921 it was clear to all interested parties that a rational, national system of forest use and conservation was no nearer than it had been in 1917. As the smoke settled on nearly a decade of war and revolution, however, and as the forest lessons of the first five years of Soviet power could be more carefully assessed, hopes rose again that a new round of legislation might achieve the still-desired goals. Forestry, in fact, was about to undergo a second transformation.

---

[86] Ibid.

VII

1921-1923 and Beyond:

The Decline of the "Basic Law on Forests" and the Creation and Impact of the 1923 "Forest Code"

Given the lack of success in organizing rational forest use or protection during the first five years of so of Soviet power, it is hardly surprising that Faleev's 1918 assessment of "On Forests" as the pride of revolutionary foresters soon gave way to a more general verdict of it as a signal failure. If in 1918 the decree's critics were mostly peasants and SRs upset by its "excessive" centralism and numerous inconveniences, by the early 1920s the Bolshevik authorities, central forest apparatus included, had joined the chorus. As will be seen, these latter groups to some degree had even begun to come round to a defense of increased decentralization. An All-Russian Forest Conference in 1921 went so far as to claim that "On Forests" had brought "ruin to forest affairs."[1] Here and elsewhere it was precisely in the field of conservation that its faults were most obvious. In this area, indeed, it was a failure twice over.

First, the law had not been successful in bringing order and reason to authorized cutting by and for the state. It was pointed out in 1923, for example, that during the operating year just ended (1922-1923) over-cutting by government organs in European Russian—defined as cutting in excess of authorized amounts—had been approximately 100%.[2] A similar, but more detailed picture was presented in the Soviet Central Executive Committee the following year. This focused mainly on the activities of the *zheleskomy*, who, it turned out, more or less since their creation had been regularly cutting from 15-*verst* strips—mostly along railroad lines or rivers—an amount of trees grossly in excess both of intended limits and of what was thought to be sustainable even in a strip twice as wide. During 1918-1919, for example, of a "permissible" 10,400,000 "kuby"[3]

[1] "Osnovnoi zakon o lesakh," *Lesopromyshlennoe delo* 3-4 (1922): 15.

[2] RSFSR, *Vtoraia sessiia vserossiiskogo tsentral'nogo ispolnitel'nogo komiteta.* X sozyva. Stenograficheskii otchet. Zasedanie vtoroe (30 June, 1923), 15-16 and passim.

[3] Again, it is not clear whether cubic meters or cubic *sazhens* are intended. On the other hand, since

14,500,000 had been cut. In 1920 of 10,192,000, 22,029,000 had gone. In 1920-1921, the intended limit had been 10 million, the actual figure 22,590,000.[4] Reckoned against central forest apparatus "targets" (apparently lower than the previously noted "authorized amounts") this meant that particularly along railroad lines in parts of European Russia cutting had by 1922 already gone "10-15-20 years ahead" of schedule.[5] The "immense amount" of forest lost just in this fashion (and subsequently turned over to peasant agriculture) was estimated in early 1923 by *Narkomzem* at 243,545 *desiatina*s.[6] Even in less heavily exploited areas in the far north, the *zheleskomy* were bitterly criticized for their forest activities on behalf of the state. The Karelian Soviet, for example, meeting in 1921, complained about these organs' "increasingly unplanned cutting"[7] and accused them of "predatory destruction" of local forests.[8]

Second, the law had also made little impact on *un*authorized cutting, which by all accounts had by 1921 slowed only slightly (and even then in some cases only because there was little of much value left to plunder). "On Forests'" impotence could, of course, have been judged primarily a failure of implementation or enforcement. While this was the case to some degree regarding authorized cutting, it was not so regarding unauthorized damage. Here, more and more, blame was laid specifically on the system enshrined in the law itself. In

---

what is important here is the discrepancy between planned and actual figures, the unit of measurement is not critically important. See also chapter 6.

[4] RSFSR. *Vserossiiskii tsentral'nyi ispolnitel'nyi komitet.* XI sozyv. Tret'ia sessiia. Zasedanie vos'moe (15 October, 1924), 349.

[5] Ibid. Another estimate—of cutting having gone "7-10 years ahead" of schedule—is given in ibid., 367.

[6] RSFSR, *Vtoraia sessiia vserossiiskogo tsentral'nogo ispolnitel'nogo komiteta.* X sozyva. Stenograficheskii otchet. Zasedanie vtoroe (30 June, 1923), 28.

[7] *Vsekarel'skii s"ezd predstavitelei trudiashchikhsia karel, 1-3 iuliia 1920* [and] *Vsekarel'skii s"ezd sovetov rabochykh, krestianskykh i krasnoarmeiskykh deputatov, 11-18 fevralia 1921. Protokoly* (Petrozavodsk: "Karelia," 1990), 211. The quotation is from the second of these two congresses.

[8] Ibid., 130. By 1921 there were 23 *zheleskomy* (and 72 *gubleskomy*) organized under *Glavleskom*, VSNKh. See: RSFSR, *Tsentral'nyi Ispolnitel'nyi Komitet. I-IV sessii vserossiiskogo tsentral'nogo ispolnitel'nogo komiteta.* VIII sozyva. Stenograficheskii otchet (Gosudarstvennoe izdatel'stvo, 1922). II sessiia, 61, 63, passim.

particular, concerns were voiced that the early push for a single plan on a national scale had presented insurmountable—or at least highly agitating—problems for the peasantry, forcing them, for example, to travel unreasonable distances either to cut wood in designated areas or to obtain already-processed supplies from officially-designated stockpiles. (These often existed only on paper.) Thus, rather than comply with rules that appeared burdensome or pointless, peasants continued simply to ignore them, to the detriment of the forests. Better forest and timber assignation, it was widely felt, would be necessary to reduce unauthorized cutting.

Popular resentment of the "Basic Law" was mainly manifested simple refusal to abide by it, that is, by the epidemic of illegal cutting. Less publicized, but certainly no less serious was the declining respect among segments of the population for forest guards, especially that apparent minority who tried to enforce the law. Statistical evidence for violence against forest guards is patchy at best, but still offers useful insights. During 1923-1924, which seems to have been the worst year, eighty-eight murders of forest personnel were reported to the Central Committee of the Forest Administration. This included two *lesnichye* (educated foresters of high rank), one assistant *lesnichii*, twenty-five *ob"ezdniki*, and sixty *lesniki* (the latter two were ranks of forest guards or rangers lacking higher education). While it cannot be ascertained what precisely were the causes of these incidents, the conclusion drawn in the journal reporting these figures was that they reflected the hostility of segments of the rural population to Soviet forest law in general and in particular a recent increase in enforcement.[9]

These problems notwithstanding, as the Civil War headed into its final stages in 1921, and as the condition of the forests touched new lows, the state's will to protect them in fact showed signs of renewed vigor. Thus, with the worst of the fuel crisis now over—and after several years of extremely heavy cutting along railroad lines—the Soviet of Labor and Defense[10] (*Sovet truda i oborony*,

---

[9] "Zverskoi ubiistvo 19 chelovek," *Lesovod* 1 (1924), 70. The figures contain an error in the original source. The numbers of murders for each category of forest workers actually add up to 88 (the number I give in the text), but are totaled in *Lesovod* incorrectly at 78. It is possible the figure of 78 is correct and there is a misprint in the list of numbers making up this total.

[10] A commission within the SNK created in April 1920 by reorganization of the former Soviet of Peasant and Worker Defense. The STO's main concern was to unite various organs' work in defense of the nation and of socialist economic construction.

hereafter: STO) made an effort in March 1921, with Lenin's involvement, to work out plans to replant nationwide alongside the tracks themselves.[11] The following month it instructed *Glavleskom*, apparently for the first time, to "curtail cutting for state needs in arid and poorly-forested areas," to "limit forest cutting by headwaters [*v verkhov'iakh*] and along the shores of rivers," and to transfer these activities to more suitable areas.[12] Also in April, the STO ordered the Commissariat of Transport to conserve fuel (wood and other sources) including by re-burning incompletely spent fuel where possible.[13] The following March, in order to minimize conditions promoting forest fires and harmful insect infestations, the decree "On Cleaning Up Cutting Debris in Areas of Forest Processing" established deadlines and standards for post-felling clean-ups.[14]

Yet these and other measures did nothing to revive the flagging reputation of "On Forests" or hopes for its proper implementation. Interested parties spoke now of the need to "begin again" with a new forest constitution enshrining a new approach overall. Above all, there was a growing sense that relations needed to be improved between the peasantry and central forest apparatus and that toward this end some kind of compromise with rural sentiment was inevitable. In the aftermath of the Civil War, moreover, the ongoing "war between the population and the forest economy"[15] could no longer be viewed in isolation, but appeared

---

[11] "Postanovlenie STO o sozdanii lesozashchitnykh polos vdol' zheleznykh dorog po vsei seti Respubliki," *Dekrety Sovetskoi vlasti*, 13: 409. The results of this initiative cannot be determined here.

[12] "O bor'be s zasukhoi," *Sobranie uzakonenii i rasporiazhenii raboche-krestianskogo pravitel'stva* (1921), 49: 250.

[13] "Postanovlenie STO o merakh po ekonomii topliva," *Dekrety Sovetskoi vlasti*, 8: 309-12.

[14] "Ob ochistke ploshchadei lesorazrabotok ot porubochnykh ostatkov," *Sobranie uzakonenii i rasporiazhenii raboche-krestianskogo pravitel'stva* (1922), 26: 307. On the other hand, timber supply remained a major priority, only slowly and slightly to diminish as other fuels became more available over the coming months and years. Organs and priorities created during the fuel crisis continued in effect with only minor alterations. At the same time, efforts to re-establish an export market for timber placed new pressures on the forests, though unlike those connected with the fuel crisis, these were more compatible with long-term planning and a broadening of the area of cutting to include under-utilized resources in the Far North and Siberia. Opportunities thus existed for integrating protection and exploitation.

[15] RSFSR, *Vserossiiskii tsentral'nyi ispolnitel'nyi komitet*. XI sozyv. Tret'ia sessiia. Zasedanie vos'moe (15 October, 1924), 368.

more and more as simply one facet of broad popular dissatisfaction with Bolshevik government. Peasant resentment at years of grain requisitioning (a central feature of the policy of War Communism) had reached the boiling point and needed to be addressed, for example. More ominously, in March 1921 workers at the Kronstadt naval base in Leningrad—who only four short years earlier had mutinied in support of the Bolsheviks—now rose up against them. More perhaps than any other single event this latter one convinced the Bolshevik regime of the need for a more conciliatory approach to its relations with major segments of the population.

In an effort to heal these kinds of tensions—as well as to restore a shattered economy—the Bolshevik government that spring effected a fundamental general policy shift, away from War Communism and into the New Economic Policy (NEP). Where the former had stifled private initiative and imposed a large measure of state control throughout the economy, the latter was to allow considerable market activity. Grain requisitions were ended (to be replaced by a tax) and peasants were now allowed to sell their surplus on the market, for example. And while the economic "commanding heights"—banking, shipping, railroads, and so on—remained government-owned and managed, enterprises employing fewer than twenty persons gained considerable freedom of action within the private sector.

Forestry, however, was at first only slightly affected by these changes (such as by the creation of "concessionary" forest use, see further below). Precisely because of this lag, however, "On Forests," so clearly formed for an era of War Communism, began in the early days of the NEP to seem increasingly anachronistic. By spring 1923 it had become, in the opinion of several members of the Soviet Central Executive Committee, "completely outdated" and "fundamentally at odds with recent central decrees" underpinning the NEP.[16] Forest nationalization carried out under its auspices was now judged to have been done excessively "straightforwardly," directly causing resentment among local populations while simultaneously failing to meet local demands for forest materials. The time could no longer be delayed to bring the NEP to forestry. Like the Kronstadt rebellion and resistance to earlier grain requisitioning, if on a somewhat lesser scale, the peasantry's forest-use complaints also represented a

[16] RSFSR, *Vtoraia sessiia vserossiiskogo tsentral'nogo ispolnitel'nogo komiteta.* X sozyva. Stenograficheskii otchet. Zasedanie vtoroe (30 June, 1923), 15.

challenge to Soviet power that could not safely be ignored. "At every village meeting," Kozyrev noted in the Soviet Central Executive Committee in 1923, "there is a general and unanimous howl . . . 'Give us the opportunity at least to make a horse-harness [by cutting down wood] near the village, rather than forcing us to go to the town to get one."[17] These kinds of popular complaints were, in fact, only to intensify over the next few years, even as the central government tried to address them.

The 1923 "Forest Code" (*Lesnoi kodeks*): Background and Significance

The substance of a new forest constitution—to be issued 7 July, 1923 as the "Forest Code"—had in fact been proposed as early as 1921 at the Tenth Congress of Soviets. It hinged primarily on a proposal to separate out state and peasant—and thus nationally-planned and locally-exploited—forest domains. To this end, and in order "to meet the wishes of the peasantry and in order to facilitate their [efforts to] raise the agricultural economy" it had at this time been accepted as "necessary to re-examine the republic's forest holdings for the purpose of establishing forests of local significance to be given over for use by the agricultural [*zemledel'cheskoe*] population under the control and to be run by the plan of local agricultural organs."[18] The authorities thus hoped that by identifying areas of forest in relatively heavily populated areas which were of need to local populations, and by then turning them over to the latter's exploitation, matters would be greatly improved: the state and its regulations would no longer be omnipresent in village life; state-rural tensions would diminish; and peasants could provide for their needs in a legal and more regulated fashion. At the same time, by delegating responsibilities to local management, resources would be freed up allowing the state to move more quickly to inventory, open up to rational exploitation, and also to protect larger and more economically important forest expanses including beyond European Russia. In plain language, then, the new law was to be a step back from rational—and especially from national—forestry, and a concession to peasant and local forest interests. Whether it was envisioned as a temporary measure pending the effect of continued efforts at "education," or as a more permanent compromise, is unclear

---

17 Ibid.

18 Ibid.

(this, of course, is the case with the NEP in general).

Either way, the first steps in the direction of reconciliation with the peasantry on forest matters had already been undertaken. A decree of 4 June 1919, for example, issued at the height of the fuel crisis, had instructed organs of *Narkomzem* and the VSNKh to waive all "special formalities" for timber procurement and to concentrate instead on satisfying popular needs—for "rough timber and dead wood, and for all types of scrap wood (left-overs from cutting, from the creation of other through-routes [*prorezhivanye*], and so on) . . . on the simplest possible bases" (article 3). It also slackened "On Forests'" requirements against cattle-grazing, allowing this now to proceed with oversight only by the peasants themselves (article 7).[19] In all other ways, however, the system of national planning in accordance with central directives had at this time remained in place in law, even if much less so in practice. By 1921, however, the decision had obviously been made to move law more into accordance with practice rather than to struggle any more to bring practice into line with paper plans—thus the need to replace "On Forests."

No attempt will be made here to provide a detailed, step-by-step account of each draft, debate, and rescript upon the "Forest Code's" path from concept to published decree, except to note certain differences between the process this time around compared with the creation of "On Forests." For one thing, the "Forest Code" does not appear to bear the stamp of any one person in the way "On Forests" does of Faleev. Available information suggests a more collaborative effort by members of the central forest apparatus and other parties within *Narkomzem* with the Bolshevik leadership based on a general consensus formed within the framework of the NEP and focused primarily on improving the conditions of the peasantry. With the SRs long ago suppressed and the Foresters' Union out of business by 1921[20] there was also little in the way of obvious input from peasant representatives or other political "outsiders"—certainly nothing comparable to the many stinging remarks endured by Faleev either from the leadership of the Foresters' Union or peasant representatives back in 1918.

---

[19] "Postanovlenie VTsIK i SNK ob otpuske trudovomu naseleniiu lesa i o pol'zovaniiakh v lesakh," *Dekrety Sovetskoi vlasti*, 5: 267-69.

[20] The Foresters' Union was dissolved in late 1921. I have not been able to find any specific information on the circumstances of this.

Indeed, in contrast with the boisterous and lively environment surrounding the creation of "On Forests," the writing of the "Forest Code" appears to have been fairly mundane.

In line with the pro-peasant and decentralizing tendencies of the day, drafts of the new law placed "particular emphasis" from the start on the "utilization of forests for the needs of agriculture [*zemlepol'zovanie*]." This was to include, for example, increased provision for forest clearing, especially where deemed "necessary for the creation of state farms [*sovkhozy*] or peasant communities [*poselky*]."[21] The new pro-agricultural line was reflected also in an increase of the authority of land organs over forest affairs, prompting a comment during discussion of one of the drafts that, "of forest organs . . . virtually nothing was noted and everything was concentrated [instead] in the hands of the land organs."[22] As will be seen, this was somewhat the case with the final decree as well.

In writing and revising early drafts a "bottom-up approach" seems to have been used. Much of the groundwork—in the form of a rough draft—was completed in Moscow during 1921 at the First All-Russian Forest Conference, attended by a variety of forest specialists and officials. This was then passed up to *Tsentroles*, now usually referred to as the "Central Forest Administration" (and so named hereafter) within *Narkomzem*. Beyond this, however, it passed out of the immediate control of the Central Forest Administration into the hands of its agricultural parent board, the *Kollegia zemledeliia* of *Narkomzem*, whose prime concern, as noted, was to ensure a smooth fit between the new forest law and overall "agrarian policy." Here it remained when final debates and rescripts were undertaken in the spring of 1923.

By this time, the emphasis on agriculture was uniting in criticism of the new law the traditionally mutually suspicious advocates and representatives of forest economics or *lesokhoziaistvo* (the Central Forest Administration of *Narkomzem*) and those of forest industry or *lesopromyshlennost'* (*Glavleskom* and its parent, the VSNKh).[23] The former resented the lack of attention given to strictly silvicultural (as opposed to agricultural) matters. The latter complained

[21] V. Maier, "Lesnoi Kodeks," *Lesopromyshlennoe delo* 13-14 (1923): 1. This refers to article 78, note [*primechanie*] "b" of the version accepted by the *Kollegia Narkomzema* on 25 April, 1923.

[22] Ibid.

[23] See chapter 5.

that "the needs of industry had been completely disregarded."[24] This temporary and basically coincidental alliance was short-lived, however. During a review of the draft in June, the Soviet Central Executive Committee, finding "guarantees . . . [for] the development of industry" to be lacking, "initiated a series" of further pro-industry amendments, the "most important" of which, by means of long-term forest concessions and other supply guarantees, "gave industry the opportunity to draw up long term plans for its development." This boost for the *lesopromyshlennost'* community was not to be matched with support for *lesokhoziaistvo*, however, leaving the Central Forest Administration somewhat out in the cold. This was perhaps all the more palpable since the Central Forest Administration's parent, *Narkomzem*, was able to use the pro-peasant emphasis of the new forest law to win back for itself some of the influence over forest politics (itself increasingly being redefined simply as a branch of agriculture) which it had surrendered during the days of the fuel crisis.[25] The new draft, then—and the final decree itself—combined very much the interests of agriculture and industry, and very little of forest management (including forest protection). Against this complex background of inter- and intra-departmental dynamics, the draft decree itself was provisionally passed by the VTsIK on 30 June and then finally so by the same body on 7 July. It went into effect on 1 August.[26]

The "Forest Code": An Overview[27]

As with "On Forests," the "Forest Code"'s overarching—and this time explicit—long-term goal was to plan and organize forest use so that 100% of yearly growth, no more nor less, would fall under the axe, representing a perfect and theoretically sustainable balance between production and consumption, exploitation and conservation (article 17). This was of course an ideal more than a realizable goal, and could only ever have been approached asymptotically. In

[24] V. Maier, "Lesnoi Kodeks," 1.

[25] Ibid.

[26] Ibid., 2.

[27] Unless otherwise stated, all citations from the "Forest Code" (*Lesnoi kodeks*) are from *Sobranie uzakonenii i rasporiazhenii raboche-krestianskogo pravitel'stva* (1924), 58: 564 (August 13, 1923).

general, working toward it would require not only better conservation in easily accessible and already exploited areas, but massive feats of research, inventorying, and the improvement of transportation infrastructure in order to open up the huge northern and far-eastern resources so far left relatively untouched.

In accordance with the dominance of agricultural and industrial influence in the decree's making, the preamble to the "Forest Code" made clear its utilitarian general priorities. These were to "strengthen and develop the agricultural economy" and to "place the forest economy on a firm basis, guaranteeing wood [supplies] to both the working population and forest industry." While reaffirming certain principles basic to "On Forests," such as the ownership of all forests by the state and the overall goal of a planned forest economy run for the benefit of the working population, it also differed from its predecessor in various ways, minor and major. Among the former: in the estimation of some of its promoters it was "more well-developed" and "detailed" than its "vague" and "declarative" ancestor (though this is not obvious to the modern reader); and there were some changes in the distribution of power among the various forest governmental levels—thus, where "On Forests" had accorded the "right to intervene" in forest affairs to "the greatest variety of Soviet organs" down to the lowest jurisdictional levels, the "Forest Code" simplified matters by "establish[ing] firmly" that the forests were the exclusive preserve of the "appropriate [level of] the Commissariat of Agriculture [*Narkomzem*]."[28] (This, of course, was far from true: as already outlined, several other organs, especially the VSNKh, continued to compete with *Narkomzem* on forest matters; and further reductions of the latter organ's forest authority were still to come).

These and other relatively small changes, however, were completely overshadowed by the decree's main innovation, the transfer of large amounts of forest from central to local control. To this end *Narkomzem* was obliged to "inspect and transfer to the working agricultural population, within a two-year period, forest areas of local significance and [other] land areas having no forest-economic significance . . . for the satisfaction of the needs of the . . . population" (preamble, article 3). This, again, represented a major step back from the earlier goal of a single, national forest plan. It was also a direct violation of Lenin's 1918 assertion that the forests, as a "general-public fund" (*obshchenarodnyi fond*) constitute the "property not of any village, province, or *oblast'*, . . . and cannot in

[28] P. Serebrennikov, "O novom lesnom kodekse," 9.

any case be subject to any division."[29] Forest division, however, was the essence of the new law. (Though Lenin was incapacitated by ill-health at the time the "Forest Code" was finally worked out and enacted, there is no indication he would have had any objections to these changes, which fitted well within Lenin's own New Economic Policy.) Early estimates spoke of figures in the region of 12 ½ million *desiatina*s to be given over to local populations. This figure quickly swelled, as will be seen. Almost all this land, of course, was to come out of the general state holding—and from many of the theoretical protections previously obtaining therein—to be turned over to the more unregulated[30] use of the working peasantry (article 2). Clearly, the state's earlier commitment to a unified national forestry was being either shelved or sacrificed in favor of the creation of a dual system, splitting the nation's forests into "forests of local significance"—to be run largely locally and for local needs—and "forests of general state significance" (article. 4)"—to remain under central control and planning and serving national needs.

Regarding which areas exactly were to be given over as of local significance, the "Forest Code" targeted lands which had been used or held by the peasantry prior to the War. Specifically, this meant those belonging to peasant "societies and associations" (*obshchestva i tovarishchestva*)—or, as worded in debate in the All-Russian Soviet Central Executive Committee in 1923, not only to "peasant societies," but also to "separate and individual" (*otdelnye i*

[29] *Lesnoe khoziaistvo SSSR za 50 let, 1917-1967*, 16-17.

[30] A sense of the greatly increased freedoms given peasants to exploit local forests as they saw fit is given by the following statement, made in the Soviet Central Executive Committee in 1925: "Forests [of local significance] will be [given] over in total, permanently, and free of charge to the disposal of the peasant population, and the peasantry will now be able, by carrying out its own forest economy by the simplest methods and without any unnecessary formalities, to satisfy its main needs for timber." (See: RSFSR, *Vserossiiskii tsentral'nyi ispolnitel'nyi komitet.* XII sozyv. Vtororia sessiia. Zasedanie desiatoe [23 October, 1925], 507. On the other hand, legal restrictions did exist on exploiting these forests (see text). These were significantly reduced in comparison to those obtaining in the bulk of general state forests, however, as well as in most forests pursuant to 1918's "On Forests."

edinolichnye)—and presumably peasant—owners."[31] In the second case, relevant land was estimated to cover nearly 15 million *desiatinas* distributed very unevenly across the country as shown in fig. 4.

Figure 4: Area of Land Identified for Transfer as "Forests of Local Significance" as of June 1923[32]

1. Northern region	2,797,836 *desiatinas*
2. North-eastern region	2,752,157
3. Central-industrial [region]	521,660
4. Petrograd region	94,750
5. Western region	1,213,572
6. Zavol'zhskii region	2,270,943
7. Central Volga region	970,814
8. Central agricultural region	670,982
9. Southeast region	49,738
10. Predkavkaz'e	13,312
Subtotal	**11,355,764 *desiatinas***
Scattered "inconvenient" or less-accessible land	3,500,000 (approx.)
Total	**14,855,764 (approx.)**

[31] RSFSR, *Vtoraia sessiia Vserossiiskogo tsentral'nogo ispolnitel'nogo* komiteta. X sozyva. Stenograficheskii otchet. Zasedanie vtoroe (30 June, 1923), 19.

[32] Ibid.

A statement made in interpretation of the new law in the Soviet Central Executive Committee included as eligible for transfer, in addition to those lands already noted, also "all other forest area which *could have been in peasant use* before the revolution [emphasis added]."[33] The Code itself also granted to peasant communities all "insignificant forest areas" (article 2),[34] defined elsewhere as "small parcels situated separately from forest massifs, [and] not having any general state significance."[35] In well-forested northern areas land could be given over even if it did not fall into either of these categories (article 5).

In accordance with what has already been noted about the rise of agricultural influence over forest matters, the decree also provided that forests of local significance were no longer to be administrated in accordance with forest law at all but rather on the basis of agricultural or land law (article 3). Thus, in these forests the authority of the Central Forest Administration of *Narkomzem* was reduced significantly, from executive and legislative to merely administrative. Its place was taken by a variety of land organs,[36] with primary jurisdiction to be constituted at the provincial level. From a conservation point of view, this was anything but encouraging. Peasant culture and land politics being what they were, this was not entirely unlike having the fox guard the hens. Worse, though land organs had gained at least some experience in conservation during the era of the Provisional Government, Soviet land law currently did not even mention the subject explicitly. On the other hand, the "Forest Code" did seek to mitigate this lack by establishing in local forests "simplified" management plans, to be confirmed by provisional-level forest organs (articles 8 and 9), and in accordance with which newly-empowered land organs would be made

[33] Ibid.

[34] The full text stated that forests of local significance were to include primarily "former peasant, communal [*obshchestvennye*] and resettlement [*pereselencheskye*] plots . . . forests under the possession of peasant homesteads [*v podvornom krest'ianskom vladenii*] (farms [*khutora*] . . .) [and] other forests belonging to peasant agricultural societies [*obshchestva*] and associations [*tovarishchestva*] prior to nationalization . . . [and which] had no general state significance" section 2. See n. 27 for source citation.

[35] RSFSR, *Vtoraia sessiia Vserossiiskogo tsentral'nogo ispolnitel'nogo komiteta.* X sozyva. Stenograficheskii otchet. Zasedanie vtoroe (30 June, 1923), 19.

[36] "Land Societies" (*Zemel'nye obshchestva*), "Agricultural Communes and Artels" (Sel'sko-khoziaistvennye kommuny i artely), and other "analagous" entities (article 7).

responsible for protecting their forests against fires, unauthorized cutting, inappropriate cattle-grazing, and other abuses. In most cases, however, these envisioned protections appear to have been minimal and left largely up to local initiative. The Central Forest Administration of *Narkomzem* was assigned the "right of overseeing the carrying-out"—but not of devising—these regimes.[37]

In other ways too, the "Forest Code" appeared less immediately committed to conservation than had "On Forests." In *all* forests, for example, both local and "general state" (though not in protected forests) the Code permitted permanent forest clearing for any of the following purposes: the needs of railroads, towpaths (*bechevniky*), for creating training grounds and firing-ranges (*polygony i strel'bishy*), camps, buildings, and for unspecified other "necessities of state significance," also when "unavoidably necessary" for improving the agricultural economy, and, finally, for the purposes of colonization (this last, unlike the rest, required a "special legislative act," see article 59). In areas where forest cover was less than 35% of total land, however, cutting—even for these purposes—needed central authorization. Central permission was legally required also in more heavily forested regions where cuts "in each separate case" were to be of more than fifty *desiatina*s. This seems a dubious deterrent to larger clear-cuts, however, since one could, presumably, legally clear much larger areas simply by doing so in fifty-*desiatina* bites. In all other cases, decisions to cut on the above bases were to be taken at the provincial level. (article 60). And still more cutting could be authorized at any time by special decree by *Narkomzem* (article 12).[38]

While local forests formed a single general category, those of "general state significance" were divided again in two: forests "properly belonging to the state" (*lesa sobstvenno gosudarstvennye*), and forests "of special significance"

---

[37] P. Serebrennikov, "O novom lesnom kodekse," 10. There is some evidence of efforts having been made during the drafting of the "Forest Code" to introduce further guarantees of some level of conservation. An unspecified early draft, for example, apparently had granted local agricultural societies [*sel'skye obshchestva*] "the right to impose specific penalties on persons found [carrying out] predatory destruction of forests of local significance." This clause did not make it into the final Code, however, which dealt with matters of enforcement and punishment in rather vague language, making confirmed acts of "improper conduct" of the forest economy in general "criminal actions" punishable under criminal law. See RSFSR, *Vtoraia sessiia Vserossiiskogo tsentral'nogo ispolnitel'nogo komiteta*. X sozyva. Stenograficheskii otchet. Zasedanie vtoroe (30 June, 1923), 20.

[38] See also ibid., 20-21.

(article 14). The former—the overwhelming majority—were to be the focus of the national forest economy primarily as sources of timber. The latter included "protected forests," "educational-experimental" forests, and monuments of nature, [as well as] mining [*gornozavodskye*] forests, urban woods, concessionary forests [see below]," and those assigned "for the specific use of transportation or industrial enterprises utilizing combined economies [*v poriadke kombinirovannykh khoziaistv*]" (article 38). Protected forests, like all forests of special designation, could be returned to the status of general state holdings essentially at will and without notice—for example, when the "purpose" for which they had originally been assigned of special designation had "passed" (article 40).

"Protected forests" were here explained in terms little altered from "On Forests" as those protecting water sources, preventing soil erosion or the silting up of waterways, and so forth, and also as forests "having hygienic or aesthetic significance" or as "monuments of nature . . . subject to preservation in their natural condition on some scientific basis" (article 41). These last, it was provided, could "in special cases" be transferred out of the control of *Narkomzem* to "other departments" (article 44). The original creation of a protected forest, however, in contrast with "On Forests" was no longer worded as the result of popular or local petition, but only as requiring "special rulings" by *Narkomzem* and the STO (article 42).

The act of stepping back from a single, national forest—and from a single, national plan—in fact went beyond even the excision of large amounts of land as local forests. Within the forests of general state significance numerous further, more specific exclusions and exceptions from the overall regime were also envisioned. For example, according to article 47, major industries could be assigned exclusive use rights of forests whose yearly growth was sufficient to cover their needs. These would then be exploited not in accordance with a single central plan but case by case in accordance with "special agreements" made among *Narkomzem* (not, significantly, its subordinate Central Forest Administration), the *VSNKh*, and *NKPS* (*Narodnyi komitet putei soobshcheniia*, the chief transport organ), but in every case to be confirmed by the STO (article 48). "Urban forests," also a subcategory of general state forests were to be put under control of local urban soviets who could then decide themselves to declare them fully protected or partially so[39] (article 51).

More significant in terms of further reducing the simplicity and unity of forest organization and planning, the new law also sanctioned and codified the growing practice of "concessionary forestry." Clearly reflecting NEP priorities and methods, concessionary forestry was intended above all to "attract . . . foreign capital" by allowing private interests to carry out logging operations on Soviet territory.[40] The "Forest Code" granted concession holders exclusive use of specific forest areas, terms of use to be set by "special agreement" in each case rather than under the umbrella of general rules for other forests of general state significance. As with the forests of local significance, here too the Code sharply reduced the Central Forest Administration of *Narkomzem*'s decisive authority. Regarding the choice of areas to be conceded, the organ was to have only a consultative role. In establishing use and conservation regimes for concessions it was to have no authority whatsoever. In both these areas the SNK,[41] in the form of its Main Concessions Committee (*Glavnyi kontsessionyi komitet* or GKK, article 53), and the STO dominated (articles 54-55, 57, 63). Thus, as with local forests, so in concessionary ones, the Central Forest Administration of *Narkomzem*'s authority was limited overall to basic administration; which meant also that concessionary forests too were essentially excluded from the various regulations and protections obtaining elsewhere in the state's general forest holdings.[42]

If it made little conservation sense, from an economic point of view (and within the context of the NEP) the logic of concessionary forestry was obvious. A 1923 article by G. Piatakov published in *Lesopromyshlennoe delo* explained that

---

[39] Rendered here as "zapovednye," rather than "zashchitnye."

[40] V. Maier, "Lesnoi Kodeks," 2. The practice seems to have begun with decrees by the STO on 3 November 1922 and 31 January 1923 allowing long-term grants of forests to domestic industrial users in the Urals. These concessions were then expanded to include foreign operators, whose presence in the Soviet economy grew quickly (ibid.).

[41] SNK—*Sovet narodnykh kommissarov* or Council of People's Commissars, the Bolshevik cabinet and chief executive/legislative organ, headed by Lenin.

[42] Each concession, it was further noted, was to have properly defined borders. There should also be a clear time limit for its operation and a basic forest plan. Concession-holders were to pay a fee for their land-use rights; and checks were to be made by Soviet organs to ensure all obtaining conditions and plans were being properly followed (article 56).

by it the Soviet state "undertakes to earn profit by use of entrepreneurs [but] . . . with the absolute condition that the country obtains genuine benefit." Forest concessions, he affirmed, "would not allow the transformation of Russia into a colony or an object of exploitation by foreign capital." "We are," he continued, "prepared [to undertake] cooperative economic work with foreign capitalists, but not to allow their interference in our politics or our social life . . . . [W]e shall continue to control the 'commanding heights' [of the economy] ourselves . . . ."[43] These were, however, general principles, not points of dogma. A few exceptions, and some overall flexibility, Piatakov added, could be allowed so long as they did not present a serious threat to the framework already noted.[44] Fundamentally, it seems, the profitability of operations was the ultimate concern. Concessionary forestry was only one, if a major, face of resurgent private forestry during the NEP-"Forest Code" years.

Beyond local forests, concessions to domestic and foreign users, and urban forests, the "Forest Code" envisioned still more exemptions and exceptions from the general state forest plan. Rules for "secondary [*pobochnyi*] forest use," including cattle-grazing, haymaking, hunting and fishing, gathering of nuts, deadwood, moss, peat, sand, clay, and so on, were also to be established separately and administered in most cases at the provincial rather than the central, level (articles 33-36).

Explicit mention of forest conservation was left to the last page of the "Forest Code," where it took up five rather generalized clauses aimed more at establishing ranks among forest personnel than at establishing any new or specific conservation or protection measures. Thus, provincial-level organs and individual *lesnichye* were made responsible for ensuring protection (article 69); guards under the control of *lesnichye* were charged with general oversight (article 70). *Lesnichye* were to run patrols in local and state forests where they would look out in particular to prevent unauthorized clearing (article 71). In response to frequent and longstanding pleas, both *lesnichye* and guards were to be armed with guns.[45]

---

[43] G. Piatakov, "Kontsessionaia politika v lesnom dele," *Lesopromyshlennoe delo* 5-6 (1923): 1.

[44] Ibid.

[45] Since 1917 at least, it had often been asserted that unarmed forest guards could do little to enforce forest protection laws. Even after 1923, however, forest guards seem to have been inadequately armed. This was in part a matter of funding, but also of lingering fears about the possible outcome of arming

And in connection with arrests of violators, general investigations, seizures of stolen materials, and in calling up peasants to help fight forest fires, *lesnichye*, their assistants, and forest guards were all granted full police powers (article 72).

In fact, conservation was not quite so forgotten as a reading just of the "Forest Code" might suggest. Much of the legal detail for forest protection—including further bureaucratic reshuffling—were treated in separate decrees, one issued beforehand (in March), the other a little later (in December) the same year. Both were based on the "Forest Code" itself, and served essentially as extra clauses thereof. The first, issued jointly by the VTsIK and SNK on 20 March, 1923 created a central Section for the Conservation of the Forests of the Republic (*Otdel okhrany lesov respubliki*). Attached and subordinated to the (by now somewhat emasculated) Central Forest Administration of *Narkomzem*, this was intended to "unify and manage the affairs of forest conservation all across the Republic."[46] Subordinate local organs—Subsections for the Conservation of Provincial Forests (*Pod"otdely Okhrany Lesov Gubernii*)—were established underneath existing Forest Sections around the country, also within *Narkomzem*. The second decree, "On Conserving the Republic's Forests," issued 20 December the same year, dealt mostly with the organization, rights, and responsibilities of a reorganized Forest Guard; and in this regard expanded on but did not alter what has already been said regarding such matters in the "Forest Code."

As passed, the "Forest Code" was not without its critics, of course. Yet few if any of the unconvinced focused on conservation deficiencies, at least not in 1923. Other issues received far more attention. Many of the most outspoken critics represented the economic organs under the VSNKh, and tended to focus their comments within larger criticisms of *Narkomzem* in general (and as already outlined), thus carrying into debate on the new Code older concerns grounded in

individuals whose political reliability was often in doubt. In this latter regard it might be noted that during final debate on the "Forest Code" in June 1923 a committee was convened to examine precisely this question (among others), in the course of which some members apparently tried unsuccessfully to kill the clause due to such fears. The issue was carried, however, by those who argued firearms were necessary to make the forest guard a viable outfit, capable of enforcing the law and of providing credible protection of the forests in their charge. See RSFSR, *Vtoraia sessiia Vserossiiskogo tsentral'nogo ispolnitel'nogo komiteta.* X sozyva. Stenograficheskii otchet. Zasedanie vtoroe (30 June, 1923), 231. The gun clause was officially endorsed by the VTsIK on 5 July.

[46] As quoted in K. Kh. Feoktistov, *Polozhenie ob okhrane lesov Respubliki i instruktsiia lesnoi strazhi po okhrane prirody*. Vtoroe izdanie (Leningrad, 1924), 6.

the complex realities of interdepartmental overlaps and tensions. In this context the needs of industrial users, despite having been better served in the final decree than in early drafts, were again reckoned to have been ignored in a rush to promote the interests of the rural economy and of the peasantry.

Perhaps the majority of complaints, however, focused on the decree as excessively favorable to private interests and less so to Soviet needs.[47] This, of course, was in effect a complaint not only about forest policy, but of the general thrust of the NEP, of which the "Forest Code" was in this respect and others only a small—but vital—part. Nonetheless, for those unhappy with the state's moderately pro-private-enterprise stance there were tangible and ominous signs. During the 1922-1923 operating year, immediately prior to passage of the Code, already about 33% of timber exports were in private hands. (Exports had been close to negligible prior to 1922, so it is impossible to make useful comparisons with earlier years.[48]) Related complaints were heard that concession-holders were being spoiled also, mostly by being given privileges and rights unavailable to other users including the Soviet state itself.[49]

There were also occasional familiar cries of excessive and stifling centralism, though these appear to have been concentrated in debate of earlier drafts, and are less noticeable after August 1923, the final draft itself allowing

---

[47] RSFSR, *Vtoraia sessiia Vserossiiskogo tsentral'nogo ispolnitel'nogo komiteta.* X sozyva. Stenograficheskii otchet. Zasedanie vtoroe (30 June, 1923), 28-29.

[48] In 1913, Russia's forest industry exported 463 million *puds* (about 7 ½ million metric tonnes). Post-Revolution figures were greatly reduced. The following numbers are expressed as a percentage of total exports for 1913): 1917, 11%; 1921, 3%; 1921-22 operating year, 9.2%; 1922-23 operating year, 22.4% (RSFSR, *Vserossiiskii tsentral'nyi ispolnitel'nyi komitet.* XI sozyv. Tret'ia sessiia. Zasedanie vos'moe (15 October, 1924), 363). Pre-war timber exports had constituted 35% of all exports by ruble value. Only in 1922 were the Soviets able properly to re-enter the export market. *Lesopromyshlennoe delo* 5-6 (1922), 1. By this time, the "Scandinavian countries and Finland" had used the opportunity presented by Russia's absence to expand their own export markets, making Russia's task all the more daunting. By 1922 Russia committed to winning these markets back and also to expanding into Turkestan, Persia and Turkey.

[49] See debate in RSFSR, *Vtoraia sessiia Vserossiiskogo tsentral'nogo ispolnitel'nogo komiteta.* X sozyv. Stenograficheskii otchet. Zasedanie vtoroe (30 June, 1923), 29-31.

considerable decentralization compared to "On Forests."[50] Indeed, much more numerous were those who saw the Code as providing "excessive autonomy" to local populations in "matters of the administration of the forest economy."[51] In the opinion of one critic, its decentralizing nature represented a dangerous and slippery slope toward complete forest anarchy since "forest organs here and everywhere consist of persons who, in their political attitudes, are not our supporters, to say the least."[52] Final debates on the law, and in particular the decisions of commissions convened to tackle minor controversial issues, also seem to show in general the ascendance of the advocates of decentralization. Such, for example, is the decision made in early summer 1923 to eliminate altogether a draft clause regulating the peasantry's use of local forests to gather mushrooms, berries, and so on (clause 35).[53]

Despite these complaints—and perhaps more importantly, despite also the numerous and significant alterations made by the "Forest Code" to the originally-envisioned unity of Soviet forest organization—there was in 1923 considerable enthusiasm among official forestry circles that the new law might finally mark the beginning of a real and lasting reform. Certainly, there were many statements that year to the effect that the huge damage done both to Russian forests and to forestry by nearly a decade of war and Revolution, while lamentable in the extreme, was a *fait accomplis* and should thus be accepted and moved past. Some commentators, indeed, even began to speak in quasi-romantic terms of the forests not as victims of a senseless brutality, but as a bruised and bloodied but ultimately heroic ally in the victorious struggle for a new political and social order. G. Piatakov's comments that year in *Lesopromyshlennoe delo* were among several of this nature: "Forest cutting . . . and [timber] transportation during the years of the great historical battle of the classes," he claimed, had provided a

---

[50] See for example, the criticisms made by Larin in ibid., 34.

[51] Ibid., 36.

[52] The reference here was not to provincial-level executive committees (*gubispolkomy*), who enjoyed considerable power over forest matters pursuant to the new Code, but to forest organs jurisdictionally beneath these, which many saw as almost entirely insubordinate (ibid.).

[53] RSFSR, *Vtoraia sessiia Vserossiiskogo tsentral'nogo ispolnitel'nogo komiteta.* X sozyva. Stenograficheskii otchet. Zasedanie shestoe (5 July, 1923), 229. Such forest use appears in the final decree to have been left to the decision of local organs.

"historic service," and "saved our country from destruction."[54] In any case, if the recent past had been filled with disappointment and struggle, important lessons had been learned, too. Moreover, a second chance at Soviet forest organization had been won and was about to be tried. If the lofty goals of rational, national forestry had proved unworkable for now, more realistic and achievable goals had been set. Enshrined in the "Forest Code" these revolved around a more conciliatory approach to the peasantry and a more eclectic and flexible approach to forestry overall, reflected above all in the division of the forests into state and local holdings (with all the complexities and exceptions thereto already noted). Even the earlier tide of peasant unauthorized cutting had by 1923 abated at least somewhat. The worst, then, seemed to be behind.

Bolshevik Forestry, Take Two: The "Forest Code" in Practice

The "Forest Code" was launched with only a little less fanfare than its predecessor. Despite a few muted grumblings, for the most part talk was again of a new beginning, of "new horizons" opening for Russia's forests, which would afford them after an earlier false start the "opportunity to rest and recover and then to fulfil their purpose in the forest market, the agricultural economy, our factory and transport industries, and in general to reinvigorate the economic life of our country."[55] The new law was accompanied by a "three-year plan"—essentially a timetable for its implementation—which reiterated some of the main points or underlying concerns of the "Forest Code": to ensure that forest use would be planned and organized, not arbitrary; that state and authorized public demand for forest materials would be fully and consistently met; and to "extract the highest degree of profit from the forests."[56]

Based, no doubt, both on the political priority of reducing tensions with the peasantry, and on the economic imperative of improving timber supply to local populations, the establishment of local forests was given top priority. A two-year timetable was quickly established in 1923, and, as noted, some 15 million

[54] G. Piatakov, "Osnovnye zadachi lesnoi promyshlennosti," *Lesopromyshlennoe delo* 1-2 (1923), 1.

[55] "Lesnoi kodeks," 9.

[56] *Lesnoe khoziaistvo SSSR za 50 let, 1917-1967*, 18.

desiatinas of land were slated for transfer. (To give some sense of scale, this was estimated at the time to be about "five times . . . the [total] forest area of the Ukraine, and three million *desiatinas* more than the entire forest area of Germany."[57]) Work began slowly, with inventorying and organizational tasks providing the early focus. By the start of fall 1924, slightly over a year into the project, by one estimate only about 5 million *desiatinas* (about 5 ½ million hectares) had been reassigned,[58] making the work 33% complete in 50% of the scheduled time. Things seem to have sped up the following year, however; and the initial desired transfer of 15 million *desiatinas* (itself up from the 12 1/2 million figure first bruited in summer 1923) was apparently surpassed. Thus by 1925, some 16 million *desiatinas* had either been transferred already or were well along[59] (completion appears not have occurred until 1927—by which time over 17 million *desiatinas* may have been designated "local," see below).

Problems accompanied the process throughout. For one thing, large amounts of forested land could not easily be established in any particular category. Especially relevant here are forests which by virtue of their proximity to villages, previous peasant proprietorship, or for other reasons, had been slated for re-designation as forests of local significance, but which had at the same time already been officially established as "protected" under Soviet law. Typically, this was due to their importance in maintaining soil integrity along rivers and in other sensitive areas. As of 1923, at least 500,000 *desiatinas* were in this category.[60] The state, moreover, seems to have been unable to let one or other designation go, preferring instead to attempt a compromised version of both. In this spirit, two "solutions" were implemented. In a minority of cases some retention was envisioned of centrally-mandated protections and rational use requirements. More often, however, a simple substitution was made, with peasants being given other land instead so as to preserve the status of protected areas. Both these approaches,

---

[57] RSFSR, *Vserossiiskii tsentral'nyi ispolnitel'nyi komitet.* XI sozyv. Tret'ia sessiia. Zasedanie vos'moe (15 October, 1924), 365.

[58] Ibid.

[59] RSFSR, *Vserossiiskii tsentral'nyi ispolnitel'nyi komitet.* XII sozyva. Vtoraia sessiia. Zasedanie desiatoe (23 October, 1925), 513. The word "completion" should not be taken too literally here. By its very nature the work was open-ended.

[60] As "protected" areas, these fell into the category of "general state forests"; as areas previously belonging to the peasantry, they were also "forests of local significance." Ibid., 507-08.

of course, maintained also most of the inconveniences peasants had so hated, provoking further angry outbursts. Even outside these officially "protected forests," Soviet wishes to hold on at least to some semblance of planned use—even if on a lesser scale and through the medium of local land organs—also collided with peasant hopes for complete freedom of action. Thus, in a great many instances peasants still did not get stretches of forest—or the rights therein—they felt they had a fair claim to. This was especially so in the black-earth regions where forests were relatively small and timber already scarce. In well-forested northern regions matters were considerably easier, though not always ideal. And there were other problems, too. Villages did not quarrel only with Soviet authority, but in some cases with each other over rights to specific areas, often resuscitating much older squabbles.[61] Overall, in fact, despite ending up (according to official estimates) with approximately *double* the forest area they had had prior to the war,[62] peasants in many areas remained unimpressed.

Nor can all this be blamed easily on the unreasonable demands of an irrational or greedy peasantry (though such charges were sometimes levelled[63]). Real shortages continued, despite all these concessions, especially in less well-forested central agricultural regions. Debate in the Soviet Central Executive Committee during 1925 revealed, for example, that 75% of all forest areas given or intended for transfer to local needs were mere "bushes."[64] Others, including several near villages in Tver province, for example, were "swamp."[65] To be sure, much of this was the result of earlier—and in some cases, ongoing—depredations by local populations. (Ironically, it was also a reflection of the very uneven distribution of forest resources across Russia which the "intrusive" nationally-planned forest policy enshrined in "On Forests" had originally been designed to overcome.) Nonetheless, it is hardly surprising that even in late 1925, with the transfers of forests to local use significantly completed, *at best* only 60% of local

[61] Ibid., 507-40, passim.

[62] Ibid., 514.

[63] P. Morozov, "Chto delat'?," 8.

[64] RSFSR, *Vserossiiskii tsentral'nyi ispolnitel'nyi komitet*. XII sozyva. Vtoraia sessiia. Zasedanie desiatoe (23 October, 1925), 528.

[65] Ibid., 535.

timber demand, it was foreseen, could ever be met from these sources.[66] One assessment, published in the forest journal *Lesovod* in 1924 put the amount of timber that local forests could supply to local populations in heavily-populated regions as low as 14% of total demand![67] Endemic overcutting in accessible areas, then, seemed likely to remain a fixture of Russian forest life regardless of law or plan.

Whatever its bases or merits—and no matter how exasperating—peasant discontent could not simply be ignored, however. As a minority leadership of urban socialists in a vast land of "petty-bourgeois" peasants, the Bolsheviks were still very dependent at least on the acquiescence, if not the active support, of the rural masses (the Civil War having by now been won). The ongoing situation in the forests, however, remained ominously tense. It was precisely in 1924, for example, that one Bolshevik (by the name of Liutin) had been inspired to speak of a "war between the population and the forest economy."[68] Thinking, perhaps, in terms not unlike the old Chinese proverb that equates the ruler with the boat and the masses with the water (the latter can either support or capsize the former), the Bolshevik government found itself constrained in forest matters to pursue the path of compromise. And nowhere was this reality—and the Bolsheviks' awareness of it—more in evidence than during debate on forest politics in the Soviet Central Executive Committee during 1925.

In this forum, speaker after speaker, focusing solely on forest politics, told of the animosity peasants felt for Soviet power. One, "Comrade Larkin," related the gist of a conversation he had recently had, as a "representative of the central authority" with "peasants" in a timber factory in Kolomenskii *uezd* near Moscow: "What are you all doing there [in Moscow]?"—Larkin had been asked by one. To this rhetorical question the peasant had supplied his own answer: "Sitting around issuing decrees, throwing out slogans . . . ." This particular individual had then gone on to offer the following complaint, entirely typical of literally dozens of others at the time, and directed against continued, if reduced, efforts by the state

---

[66] Ibid., 521. The same figure is given in various sources. See, for example, P. Morozov, "Chto delat'?," 8.

[67] V. Solov'ev, "Les i krest'ianstvo," *Lesovod* 2-3 (1924), 13.

[68] RSFSR, *Vserossiiskii tsentral'nyi ispolnitel'nyi komitet*, XI sozyv. Tret'ia sessiia. Zasedanie vos'moe (15 October, 1924), 368.

to regulate cutting:

> "[Y]esterday there was a meeting . . . [some of our local peasants] went to the *lesnichii*, asked him to assign . . . forest for cutting firewood or [timber] for building . . . . He gave us forest land alright, but fifteen *verst*s away, when there is forest three *verst*s away. [At the same time] a village fifteen *verst*s away got given forest three *verst*s from us."[69]

Larkin used the occasion to call on *Narkomzem* then and there "to issue an order . . . that the forest be allocated near the villages, *so that there will be no censure of soviet authority* [emphasis added]."[70] Another speaker, Vasendin, pointed to the potentially "enormous . . . political . . . minuses" of not addressing the "dissatisfactions of the peasantry" in this regard.[71] A third, Terekhova, asked pointedly, "Why push the peasantry into an ill-willed relationship with Soviet power? Why not give them the forests they want, let them cut near home instead of some distance away?"[72] Still, not everyone in the Committee felt so inclined to capitulate. One speaker, Ageev, called a colleague's comments in favor of greater peasant rights in the forest "agitation designed to win over the sympathies of the peasantry." Significantly, the accused in this specific case, one Dikarev, replied, "from a political point of view, perhaps [you are] right."[73]

In response to these and other issues, and despite the Soviet government's insistence on maintaining at least some minimal level of centrally-organized

---

[69] RSFSR, *Vserossiiskii tsentral'nyi ispolnitel'nyi komitet.* XII sozyva. Vtoraia sessiia. Zasedanie desiatoe (23 October, 1925), 527.

[70] Ibid., 527-28.

[71] Ibid., 532.

[72] Ibid., 534. The threat to Bolshevik rule presented by peasant discontent had been understood in a forest context as early as 1918. In debate early that year regarding "On Forests" a speaker in the Soviet Central Executive Committee noted of burdensome forest regulations in general (especially excessive taxes), "these kinds of little mistakes can really cause offence: based on them [peasants] carry on agitation against the Soviets. Such agitation is going on and will continue based on such irritants." See *Protokoly zasedanii vserossiiskogo tsentral'nogo ispolnitel'nogo komiteta.* 4-go sozyva. Stenograficheskii otchet (1920), 319.

[73] RSFSR, *Vserossiiskii tsentral'nyi ispolnitel'nyi komitet.* XII sozyv. Vtororia sessiia. Zasedanie desiamoe (23 October, 1925), 537.

forestry, the general dynamic throughout the NEP appears to have been escalating compromise by the state with local feelings. Not only was the total amount of land to be given over revised upward several times (from 12 ½ million *desiatina*s to 15 million, then to 16 million, and ultimately to about 17 ½ million[74], but limitations on use of local forests were also relaxed. Then there were tax breaks, extra-budget distributions of timber, and so on.

If willingness to compromise with local populations derived from fears of the consequences of incurring the peasants' wrath, there seems also to have been some genuine sympathy for them, too. Rural resentment continued high, many agreed, because the basic stimulus to unauthorized cutting—the state's inability to provide adequate and convenient timber supply to local populations by means of planned forest use—was little changed. That "On Forests," had made criminals of many peasants who were simply providing for their families' basic needs was conceded by Party members on numerous occasions, such as by Comrade Izvekov, who spoke on this matter to the Central Executive Committee the previous autumn (1924): "From the first day of the Revolution the forests were all nationalized and the peasants did not have even a tiny bit of forest at their own disposal, therefore the peasantry stole forest [materials]."[75]

Overall, then, based partly on sympathy for the long-suffering peasantry, but more on fear of the latter's potential for political and social disruption (as Victor Chernov had once said, "the peasantry is the real autocrat of Russia"[76]), by the mid-1920s the state's original priority to plan and oversee forest use centrally and on a national scale was increasingly being replaced by a new prime directive: satisfy and placate the peasantry. Ia. N. Brandenburgskii, a major figure of the day in the People's Commissariat of Justice,[77] made a comparable point to the Central Executive Committee in 1926 while reviewing recent alterations to the "Forest Code": "the basic principle [of forest law] boils down to this: the population must receive forest materials and rights to secondary use of the forests

---

[74] RSFSR, *Narodnyi kommissariat zemledeliia, otchet, 1924-1925* (Moscow: Novaia derevnia, 1926), 408.

[75] RSFSR, *Vserossiiskii tsentral'nyi ispolnitel'nyi komitet*. XI sozyv. Tret'ia sessiia. Zasedanie vos'moe (15 October, 1924), 369.

[76] Sir John Maynard, *The Russian Peasant and Other Studies*, 97.

[77] Brandenburgskii (1881-1951) was a member of the Collegium of the People's Commissariat of Justice and of the Legislative Commission of the Council of Peoples' Commissars from 1922-1929.

at prices lower than those [obtaining] pre-War."[78] This was simply an alternative statement of a notion by now widely accepted among the Bolshevik apparatus: for the sake of political peace, the peasantry's current conviction—that in forest matters at least the Bolsheviks were enemies more than friends—could not be allowed to stand, and certainly must not be allowed to grow.

Yet, if the state appeared to be coming to a more sympathetic understanding of the peasantry's forest needs, it seems the latter did not reciprocate, at least not regarding conservation and planning. Matters here were as bleak in the mid-1920s as they had been in 1917. The "relationship between *lesnichye* and the peasant masses," for example, was still reckoned within the Party apparatus in 1924 to be "no good at all."[79] "In Orlovskii uezd" a local functionary reported at the same time, "in the eyes of the peasant masses . . . [organized forest work] to this day stands at the very bottom of the scale [of priorities]. Such work [they say] is of no use at all."[80] In 1924 as in 1918, Party talk was still of the "need to reconcile the population to forestry,"[81] to "explain to the peasants" the essence of "worker peasant power" and its "significance for the forests," and so on.

Circumstances in the forests, consequently, remained essentially unchanged. If the ferocity of cutting had tended generally to subside since the dark days of 1917-1918 it had not done so by a large degree. One "Comrade Burlakov," for example, having inviting the Central Executive Committee in 1924 to recall the "revolutionary days, when the forest was being destroyed as fast as possible . . . when everyone tried to cut down whatever he wanted," claimed that nothing much had changed since then: "In the localities everywhere and generally the same system is operating [now] as during those Revolutionary times." In Tver province, for example, "the population, without any supervision cuts down the forest" with no response at all from the local authorities. "In the Sofonovskii forest, near the town of Giorgevsk, cutting is going on at will." Local

[78] RSFSR, *Vserossiiskii tsentral'nyi ispolnitel'nyi komitet.* XII sozyv. Tret'iia sessiia. Zasedanie shestnadtsatoe (18 November, 1926), 847.

[79] RSFSR, *Vserossiiskii tsentral'nyi ispolnitel'nyi komitet.* XI sozyv. Tret'ia sessiia. Zasedanie vos'moe (15 October, 1924), 370.

[80] Ibid.

[81] Ibid., 371.

organs here, he claimed, and in other places too, profited from, rather than resisted these illegal activities.[82] Similar stories were told that same year in a variety of forums and concerning most major regions of the country.

In 1925, with forest reassignment significantly complete, the story remained unchanged. Reports came in from the localities to the central forest apparatus of "forests given to peasant societies [*obshchestva*—under the auspices of the "Forest Code"] which in one year had cut them down completely."[83] Nor were such reports rare. Indeed, the litany of such events, complaints, and recriminations could be carried on *ad nauseum*, so numerous are they in source materials from the period. So far as actual conditions in the forest were concerned, then, the passage and early implementation of the new "Forest Code" did not appear to be a transforming event.

Surveying the period up to 1925 more generally, a frequently cited article published that year by P. Morozov interpreted both past and present in the bleakest of terms. Forests all across the Soviet Union, and not just in the most heavily-populated areas of European Russia, were "dying irretrievably." "Everywhere," Morozov continued, "a process of cruel, slow, but clear destruction is going on—all the more terrifying, he added, because it was not always immediately obvious until afterwards. "The forests of the south and the *predstep'* [edges of the steppe lands] are almost destroyed, and the region is [consequently] experiencing drought, damaged soils, and a serious timber shortage." An "enormous swath of the northern forests" were dying off as a result of fires and insects, the impact of which was increased by poor stewardship. Dying also as a consequence, he noted, was "the legend of the unbroken tract" of Russia's northern forests. While these were problems whose roots lay far in the past, and particularly in the era of "private forest ownership," Morozov expected little if any improvement in the near future. The whole corpus of existing laws, plans, and policies he dismissed as the work of "men of the office, book, and conference." Actions—not printed or spoken words—were needed. And things were even worse, Morozov argued, than many of his peers realized. Faulty methods of inventorying forest, he claimed, hid the true level of destruction,

---

[82] RSFSR, *Vserossiiskii tsentral'nyi ispolnitel'nyi komitet*. XI sozyv. Tret'ia sessiia. Zasedanie vosmoe (15 October, 1924), 379.

[83] RSFSR, *Vserossiiskii tsentral'nyi ispolnitel'nyi komitet*. XII sozyv. Vtororia sessiia. Zasedanie desiatoe (23 October, 1925) 538.

which was worse still than even the dismal statistics of the time suggested.[84] Efforts to educate the masses on the principles and importance of forest conservation had not been carried out on anything like the required scale or in the correct manner. Consequently, scant progress had been made in breaking traditional popular antipathy toward conservation, Morozov lamented. The current cadres of forest guards, Morozov charged, did not yet even *try* to stem forest conservation and management violations. They merely catalogued them. Nor did they have much choice. Increased popular hostility to "experts" and outsiders, and the reduced status of forest guards and other vital personnel in the eyes of villages made matters almost hopeless.[85] Even viewed purely in economic terms, he claimed, things were running out of control, with the amount of timber wasted during cutting rising steadily in comparison with pre-WWI models and state profit per unit area of forest falling.[86]

[84] P. Morozov, "Chto delat'?," 5.

[85] Ibid., 11.

[86] Ibid., 10.

VIII
Lenin and Forest Protection

The emphasis given in this study to group cultures—especially the conflicting ones of state and peasant—leaves little room for detailed consideration of the many individuals involved in early Soviet forest organization and conservation. An exception has been made for N. I. Faleev, who significantly shaped "On Forests," and whose personal struggles through the law's many drafts and debates reflect so well the greater tensions swirling around Revolutionary-era forest politics in general. It will be useful here to make a second exception, however, and consider in some detail the role played by Lenin. In this case, however, it is not because Lenin is obviously central to the story (it has been told so far with only occasional reference to him). Rather, Lenin needs to be considered primarily because of his place in subsequent Soviet historiography and mythology as a great advocate and shaper of Soviet nature conservation in general and forestry in particular.

Lenin's reputation in this respect was developed not during his life, or even shortly thereafter, but after about 1950—and particularly during the 1960s. (An important monograph on the history of Soviet nature conservation published in 1947, for example, makes only the slightest—and likely the first—mention of Lenin in this regard.[1]) This is perhaps not terribly surprising. Conservation was during Lenin's lifetime still a relatively obscure concern both within Russia and internationally. It was, consequently, in little need of public heroes. However, as issues of pollution, natural resource conservation, and environmentalism attained critical momentum and mass interest around much of the developed world in the decades following World War II—eventually moving to political center-stage in the 1970s—a need appears to have been felt in the USSR to emphasize the superior credentials of Soviet socialism in this realm too. Thus, Lenin—both as the nation's real founder and its mythical hero—was essentially reconstructed as a conservation juggernaut. (During the 1930s, by way of contrast, and during a time when official culture valued industrialization above all else, Lenin was typically

[1] The book is V. N. Makarov, *Okhrana prirody v SSSR* (Moscow: Gosusarstvennoe izadatel'stvo kul'turno-prosvetitel'noi literatury, 1947).

portrayed in the forestry press as the advocate only of production, not of conservation, *per se.*[2]) He now became a man possessed of truly amazing ecological foresight, who had worked ceaselessly and achieved great successes in the service of better environmental protection or resource management in his own time, and whose vision and methods would continue to guide Soviet nature conservation along a sure path for many decades to come. In this spirit literally dozens of short books and pamphlets, and many more magazine articles, school readers, and so on were published from the 1950s to the 1980s. Here are some of their titles: *V. I. Lenin on Forests and the Forest Economy*; *Vladimir Il'ich Lenin and Nature: A Discussion for Students*; *Nature Conservation in the First Years of Soviet Power (based on materials from Lenin's Decrees)*; *V. I. Lenin on Nature Conservation; Leninist Principles of Nature Conservation*; *Vladimir Il'ich Lenin and Nature*; *From the History of Nature Conservation in the Astrakhan Area: the Materials and Activities of V. I. Lenin and his Comrades-in-Arms.*[3]

Reading these materials one is quickly faced with an apparent paradox. On the one hand, it is freely and frequently admitted that Lenin "did not write special articles on the relationship of humans to nature,"[4] that he "did not write . . . even [any] notes on the relationship of man and nature, [or] on the principles of . . . nature protection,"[5] and so on. On the other, however, we are also told, often in the very same sources, of Lenin's "great contribution to the working out of the

---

[2] Particularly representative articles abound in issues of *Lesnoi spetsialist* during 1930.)

[3] The full Russian titles of each are, respectively: P. V. Vasil'ev, "V. I. Lenin o lesakh i lesnom khoziaistve." In *Lesokhoziaistvennaia i lesomeliorativnaia nauka v SSSR, 1917-1967* (Moscow: Lesnaia promyshlennost', 1967); *Vladimir Il'ich Lenin i priroda: Beseda dlia shkol'nikov* (Irkutsk: Irkustkaia oblastnaia stantsiia iunykh naturalistov, 1980); N. A. Gladkov, *Okhrany prirody v pervye gody Sovetskoi vlasti, po materialam leninskikh dekretov* (Moscow: Izdatel'stvo moskovskogo universiteta, 1972); I. I. Khutortsov, "V. I. Lenin ob okhrane prirody" (Maikop: Adygeiskoe otdelenie krasnodarskogo knizhnogo izdatel'stva, 1969); P. I. Zhukov, *Leninskye printsipy okhrany prirody* (Minsk: Gosudarstvennyi komitet soveta ministrov Belorusskoi SSR po okhrane prirody, 1969); Iu. N. Kurazhkovskii, *Vladimir Il'ich Lenin i priroda* (Astrakhan: Nizhe-volzhskoe knizhnoe izdatel'stvo, 1969); Iu. N. Kurazhkovskii, *Iz istorii organizatsii okhrany prirody v astrakhanskom krae: Materialy i deiatel'nosti V. I. Lenina i ego soratnikov* (Astrakhan, 1959).

[4] Vladimir Il'ich Lenin i priroda, 3.

[5] N. A. Gladkov, Okhrany prirody v pervye gody Sovetskoi vlasti, 3.

problems of the interrelation of human society and nature,"[6] that he "gave great attention to the organization of correct and expedient [*tselesoobraznyi*] use of the natural resource riches of the country,"[7] that he "ceaselessly underscored the necessity of a thrifty approach *[berezhlivoe otnoshenie*] to nature,"[8] and so on. The two positions are, of course, not necessarily incompatible. Lenin *could* have been an ardent yet inconspicuous conservationist. His reticence on the subject does impose upon his apologists a considerable burden of proof, however.

Lacking many explicit sources, the Bolshevik leader's supporters built their case on two types of evidence: anecdotal and legislative. The latter rests heavily on equating Lenin's having signed particular decrees with his having taken active part in creating and promoting them; this approach and its problems are discussed further below. First, it will be useful to treat the anecdotal evidence, which appears to be more voluminous and is interesting in its own right as a window on Soviet hagiographical techniques. For the most part, I have focused on evidence directly relevant to forests, and only occasionally on material concerning other natural resources.

Lenin as Conservationist: The Anecdotal Evidence

Perhaps the first thing one notices is that though the number of books and articles is large, the information they contain is not. The same few anecdotes and examples come up time and again. Some seem rather vague or not to prove very much of anything at all. Overall, they appear to say more about his biographers' efforts to find useful material than about Lenin himself.

Most accounts begin with Lenin's early childhood, finding there the source of what is characterized as his deep and abiding love of nature. Says one: "V. I. Lenin was born in the open spaces by the mighty Russian Volga river, where the amazing beauty of the surrounding areas constantly inspired him with a love for Russian nature."[9] His childhood home, another notes, had a "large green garden with flowerbeds" and was filled with a wide variety of great trees. Here, and by

---

[6] I. I. Khutortsov, V. I. Lenin ob okhrane prirody, 3.

[7] *Vladimir Il'ich Lenin i priroda:*, 3.

[8] P. I. Zhukov, *Leninskye printsipy okhrany prirody*, 7.

[9] I. I. Khutortsov, *V. I. Lenin ob okhrane prirody*, 5.

the nearby (smaller) river Sviaga, Lenin spent happy summers walking, playing, camping, and swimming. His family also had a summer retreat about 30 miles from Samara in the village of Alakeev. Here in particular, a common version of the legend goes, the future leader of the Bolshevik Revolution underwent some of his formative experiences, combining, significantly, his two greatest loves: experiencing nature and studying the human condition. One of Lenin's brothers, Dmitri, later recalled a common scene:

> In Alakeev we lived in a small wooden home abutting a large overgrown garden separated from the neighboring field by a ditch. In the north-west corner of the garden was 'Volodia's [i.e. Lenin's] spot'—a little wooden table and bench set in the ground; this area was so completely covered in greenery that the sun could hardly penetrate. Volodia soon beat a path around the table . . . along which he frequently paced while mulling over things he had recently read. For ten years this [spot] was for Lenin a genuine working office. His work was so systematic that I can hardly remember a morning when he did not work there."[10]

Another great joy in Lenin's young life, we are sometimes told, were the many wilderness hikes he took in these regions. This was an activity he would return to enthusiastically during periods of exile in the late 1890s and, when opportunity existed, at other times, too.[11]

Lenin is also reckoned to have been something of an expert out in the field, adept at identifying various species of bird and animal, and greatly interested in observing them. He "was highly reputed" among friends "for his ability to find birds nests," for example. Unlike many other similarly gifted youngsters, however, Lenin, "protected [the nests] and never touched [them]."[12] (Several sources attest to his love for birds.) Numerous anecdotes belabor the point that Lenin not only appreciated the great beauty of the wild, but loved nature "passionately," feeling great empathy with his fellow living creatures. Even the illness some years later of his favorite potted palm apparently caused him "great anxiety."[13]

This compassionate streak is frequently noted in accounts of Lenin's hunting trips, too. One oft-repeated tale from the 1890s tells of his companions'

---

[10] Quoted in *Vladimir Il'ich Lenin i priroda*, 4. Original source not cited.

[11] See, among many sources, Iu. N. Kurazhkovskii, *Vladimir Il'ich Lenin i priroda*, 12.

[12] Ibid., 5. The original source for this story is Lenin's wife; see N. K. Krupskaia, *Iz vospominanii o V. I. Lenine. Sbornik Vospominanii o V. I. Lenine* (Moscow, 1956), vol. 1.

[13] *Vladimir Il'ich Lenin i priroda*, 4.

surprise when, presented a clear shot at a fox, Lenin hesitated and allowed the animal to escape because he had been suddenly struck with how "beautiful she was."[14] Overall, Lenin is characterized in Soviet conservation sources as having "almost never killed animals" himself while hunting, and "never to have been a hunter in his heart."[15] Accounts of Lenin in the early Soviet *hunting* press, on the other hand—which preferred, understandably to see Lenin as a great patron of the sport—paint a slightly different picture. A hunting companion from Lenin's years of Siberian exile, for example, notes that after an early slow start, Lenin "used to shoot almost without [ever] missing."[16]

Convictions and feelings nurtured in childhood apparently resurfaced continually in Lenin's adult and political life. Almost legendary is the story of how, during the depths of the fuel crisis in 1919, he found evidence in the *Sokolniki* forests near Moscow of ongoing unauthorized and potentially damaging tree-cutting by a local factory and nearby peasants. Upon viewing freshly-cut stumps Lenin snarled: "What an outrage! They plunder and destroy a forest like this! They cut down the forest, and then what? Where will the people find recreation? To destroy is simple and easy, but when will we grow [the forest] again?"[17] Lenin immediately wrote up a local ban on cutting.[18] To this experience is attributed in some sources also the broader decree signed into law by Lenin that September forbidding cutting within 30 *verst*s of Moscow (see chapter 6), as well as another decree, April 1920's "On Conserving Green Space (Gardens, Suburban Forests and Other Green Planted Areas)."[19]

The memoirs of Lenin's relatives, friends, and acquaintances (also mostly written after World War II) also contain several attestations to his interest in and

---

[14] "V. I. Lenin na okhote," *Okhotnik: Illustrirovannyi zhurnal okhoty i rybolovstva* 5 (1928): 33. See also *Vladimir Il'ich Lenin i priroda*, 6.

[15] I. I. Khutortsov, *V. I. Lenin ob okhrane prirody*, 12.

[16] See "Vospominaniia sputnikov po okhote V. I. Lenina-Ul'ianova," *Okhotnik: Illustrirovannyi zhurnal okhoty i rybolovstva* 10 (1927): 32; and "V. I. Lenin na okhote," passim.

[17] Quoted in I. I. Khutortsov, *V. I. Lenin ob okhrane prirody*, 10.

[18] The story appears in the following sources among others: *Vladimir Il'ich Lenin i priroda*, 3; P. I. Zhukov, *Leninskye printsipy okhrany prirody*, 11; V. N. Makarov, *Okhrana prirody v SSSR*.

[19] See P. I. Zhukov, *Leninskye printsipy okhrany prirody*.

fondness for nature. V. D. Bonch-Bruevich,[20] for example, wrote in 1956 that Lenin "dreamed that our Moscow and other of our large towns would be made beautiful and green, that there would be parks and squares everywhere for the masses to walk and vacation in."[21] Lenin's wife, Krupskaya, and others offer similar tidbits.

Some stories of Lenin's nature interests appear in retrospect like tailor-made hero-myths. Such are two stories about him and trees. One notes that while hiding from the police during the summer of 1917 in the village of Razliv, Lenin helped three young boys plant an oak tree against the wishes of the boys' father (no reason is given for the father's objection). In the other, Lenin was "deeply touched" by a gift of fifteen cherry saplings given to him by a delegation of weavers in 1923. True to his nature-loving side, he made sure "that [the trees] were planted in a good place and that their delicate trunks were protected from the cold."[22] Whether or not these stories are true, they clearly have about them certain mythic elements and qualities that allow them to serve as analogies for the growth of the Soviet Union and communism in general. In both cases something is planted and grows. Part of each story is the observation that the young plants are now mature trees: the cherries bear great quantities of delicious fruit; the oak stands tall and sturdy spreading its branches ever wider. Workers and young children—potent symbols of a Soviet future—assisted in the establishment of oak and cherry, respectively. In the former story, Lenin sides with youthful rebellion against the traditional authority of the parent. (Lenin's concern that the "population" have forests to "vacation in" provides also a potentially mythic element to the *Sokolniki* forest story related further above.)

What weight should be given to these various stories and anecdotes? Some, no doubt, but not nearly so much as was assigned to them in the Soviet press. For one thing, Russia in the years of Lenin's childhood was, after all, not short of countryside, forests, and wilderness. Given the attention of a generation of

[20] V. D. Bonch-Bruevich (1873-1955) was a Russian historian, and Communist Party member from 1895. He enjoyed a long and varied career writing, publishing, and editing in Imperial and Soviet Russia, and worked closely with Lenin in many capacities, including overseeing moving the capital from Petrograd to Moscow in 1918. He was, therefore, relatively well placed to offer insights and anecdotes on the Bolshevik leader's life.

[21] Quoted in I. I. Khutortsov, *V. I. Lenin ob okhrane prirody*, 14. Original source given as "V. D. Bonch-Bruevich, 1956."

[22] I. I. Khutortsov, *V. I. Lenin ob okhrane prirody*, 17.

propagandists, the childhood of millions of Russian children (and thus many revolutionaries) could no doubt have been rendered in such idyllic terms, making Lenin's connection to the natural world more ordinary than unique. Alternatively, it is likely that a thorough search of relevant source materials could turn up just as many (similarly meaningless) anecdotes arguing that Lenin cared not a jot for nature and wild scenery. The Russian playwright and historian Edvard Radzinsky offers the following vignette, for example:

> The revolutionary Maria Essen describes a walk with Lenin in the Swiss mountains. Lenin and the young woman are standing on a mountaintop: "The view is boundless. . . the glare from the snow is intolerably bright. . . . I am in the mood for high poetry . . . about to recite Shakespeare and Byron . . . when I look at Vladimir Ilyich. He sits there deep in thought, and suddenly raps out 'Say what you like, the Mensheviks are really shitting on us!'"[23]

With a little imagination, even Stalin, the undefended villain of environmentalism in western and post-Soviet Russian literature alike -- and a man almost totally glossed over in this regard even in the most gushing pro-Soviet Russian accounts—can himself be created a fairly convincing nature lover. From exile in the frozen village of Kureiko, for example, he wrote to his second wife, Olga Alliluyeva in 1913, begging to be sent "postcards with natural scenes. . . . I'm crazy with longing for natural scenes, if only on paper."[24] Alternately, Stalin-the-pioneering-ecologist could be constructed on the basis of comments such as this, from his *Voprosy Leninisma*:

> In contradistinction to metaphysics, the dialectic views nature not as a chance aggregation of objects, as phenomena, cut off from each other, isolated from each other and not depending on each other—but as a connected, single whole, where the objects, phenomena organically connected with each other, depend on each other and are conditional upon each other.[25]

And it was Stalin, finally, who at the Seventeenth Communist Party

---

[23] Edvard Radzinsky, *Stalin: The First In-Depth Biography Based on Explosive New Documents from Russia's Secret Archives*. Trans. H. T. Willets (New York: Doubleday, 1996), 54.

[24] Robert Tucker, *Stalin as Revolutionary* (New York: Norton, 1973), 159.

[25] Stalin, *Voprosy Leninisma*, 11th edition (Moscow: Gosudarstvennoe izdatel'stvo politicheskoi literatury, 1953), 536. (Cited in V. N. Makarov, *Okhrana prirody v SSSR*, 8.)

Congress (in 1934), noted that "in order to ward off droughts in the *Zavol'zhe* region it was necessary to plant forests and forest-protective strips."[26] Well-chosen anecdotes, clearly, can be put in the service of many theses. (In the course of researching this book, I even came across enough to make pretty good conservationists out of several tsars—Nicholas I and Alexanders II and III in particular—men typically excoriated in the Soviet press for heading and representing all that was wrong with capitalist and pre-Soviet exploitation of natural resources.)

If Lenin loved nature, was kind to trees, and suffered along with sick palms, in his treatment of human beings, on the other hand, he was not always quite so tender. Some individuals who illegally damaged forests or other natural resources discovered Lenin could be extremely strict. For example, on hearing in of the unauthorized cutting of a single fir-tree in the Gorky sanatorium in Podol'sk *uezd* near Moscow, Lenin personally wrote and signed a "decision" of the STO the very same day the (14 June, 1920) calling for the arrest and one-month's confinement of E. Y. Vever, the sanatorium's director, "for causing damage to Soviet property . . . ." "Any further similar offense," Lenin continued "will incur punishment for all workers and office staff [in the sanatorium] and not the manager alone."[27] Vever, evidently, protested the decision in some fashion; but Lenin was unrelenting. He wrote again two days later: "Having once again considered my decision in regard to Comrade Vever following his further explanations, I inform . . . that my decision stands and has to be fulfilled."[28] In another example, though not related in this case to forests, on hearing of "predatory" fishing practices in the headwaters of the Don, near the Sea of Azov, and in particular of the complicity in these activities of the local river guard (*okhrana vod Donprodkoma*) Lenin wrote angrily:

---

[26] Cited in V. N. Makarov, *Okhrana prirody v SSSR*, 20.

[27] Quoted from *Lenin: Collected Works* (English). Trans. Bernard Isaacs, 3rd printing (Moscow: Progress Publishers, 1977), 42: 196-97. Secondary accounts of this "decision" are given, among other places, in I. I. Khutortsov, *V. I. Lenin ob okhrane prirody*, 9; P. I. Zhukov, *Leninskye printsipy okhrany prirody*, 10; and in *Vladimir Il'ich Lenin i* priroda, 4. Interestingly, V. E. Boreiko has ascertained that the above-cited note was in fact written by Lenin *after* Vever had already been arrested. See "Ekologicheskii Leninizm." In V. E. Boreiko, *Belye Piatna*, 136.

[28] *Lenin: Collected Works*, 44: 391.

> The manager of the . . . [guard] has been relieved from his post . . . [for his role in these matters]. This person was *only* fired from his post. We must find out where he is [now] and check *seriously* if he has been punished enough [emphases in original]. . . . We ought not just to frighten him, but also sufficiently call him to account and purge him [*pochistit'*] for this outrage."[29]

Do these episodes represent the forthright anger of a man whose beloved fellow creatures had been needlessly slaughtered? Were they, perhaps, the responses of a committed conservationist presented with serious destruction of the natural environment? Either of these interpretations could be somewhat supported. So, however, could many others. For example, several recent studies have emphasized dictatorial and even quite brutal aspects of Lenin's personality. During 1918, for example, Lenin issued an order to "[h]ang (I mean hang publicly, so the people will see it) at least a hundred *kulaks*, rich bastards, and known bloodsuckers."[30] Numerous other similar statements and episodes showing Lenin in a similar light have recently been published.[31] In choosing strict punishments for the above-noted examples of nature destruction, then, Lenin may simply have been acting in character when dealing with insubordinates. The fact that the matters at hand involved natural resources may well have been more incidental than critical. Perhaps, even, what really got Lenin's goat was not so much the destruction of specific trees or fish as the disregard shown in these instances for Soviet orders and authority (and ultimately for his own authority). Ultimately, these questions must remain open, however.

Lenin as Conservationist: The Legislative Evidence

As noted earlier, the case for Lenin as progressive conservationist is based not only on anecdotal evidence but also on the corpus of conservation decrees he

[29] P. I. Zhukov, *Leninskye printsipy okhrany prirody*, 11.

[30] Quoted in Michael Scammell, "The Price of an Idea," *The New Republic* (20 December, 1999), 34. Originally in Stephane Courtois, et al. *The Black Book of Communism Crimes, Terror, Repression* (Cambridge, MA: Harvard University Press, 1999).

[31] See, for example Dmitri Volkogonov, *Lenin: A New Biography* (The Free Press, 1994); and to a lesser extent, Robert Service, *Lenin A Political Life*, 2nd edition (Bloomington, IN: Indiana University Press, 1995).

signed. Lenin, thus, was a pioneering conservationist because, for example, he signed into law, "On Forests," which "spoke of measures to protect forests, [and] about increasing the area under forest"[32]—or because "under his signature were issued no fewer than ninety-four" decrees and regulations on nature conservation.[33] The problem with these kinds of statements is that they equate the act of signing a given law with having created it. Evidence to support such as link, however, is conspicuously lacking. Nonetheless, once signed by Lenin, "On Forests" and numerous other early Soviet forest and conservation laws became "his." (Co-signers—such as Sverdlov in the case of "On Forests" —enjoy no such cachet.)

Lenin's actual involvement in conservation legislation is very difficult to reconstruct. No doubt, he would have had *some* role. As the leading figure in chief legislative bodies of the day (the SNK and STO, for example) Lenin must have been at least somewhat informed on and even personally involved in the issues that came before them, conservation and forestry included. And that he had at least some personal interest in such matters has already been established. Moreover, it is unlikely he ever simply signed important documents without reading them or—given his nature, and the prestige and power he enjoyed among the Bolsheviks—without generally agreeing with their terms. Ultimately, however, his role in the creation of most conservation legislation is simply not known in any detail or with much certainty.

In the absence of knowledge (and in pro-Leninist hands), the tendency has been to impute critical contributions to him in every case without much concern for hard evidence. In the case of "On Forests," however—one law whose creation and passage we *can* significantly recreate—Lenin's role seems to have been very small. Indeed, the law's history was described in some detail earlier in this study without any reference to him at all. Nor was this achieved by omitting references in the source materials to his contributions. There simply are none. The point at which Lenin might most conceivably have become involved in the law was shortly before its passage, when it came under the scrutiny of the SNK, which Lenin chaired (see chapter 5). The minutes of the relevant sessions, however, are an anti-climax. "On Forests" was read to the SNK on 2 May, 1918. Discussions scheduled for the 8th and the 10th were both cancelled due to Faleev's absence. A

---

[32] *Vladimir Il'ich Lenin i priroda*,, 3.

[33] N. A. Gladkov, *Okhrana prirody v pervye gody Sovetskoi vlasti*, 4. The reference here to "ninety-four" laws is the most specific and conservative Soviet estimate I have seen.

meeting on the $11^{th}/12^{th}$ was more successful. The minutes, however, record only the contributions of Faleev, Larin (a member of the Forester's Union's governing board) and the commission designed to finalize contended points (see chapter 5). The last substantive meeting, held on the 14th in Faleev's absence, formally approved the final draft. There was another related meeting on the 28th. Lenin was present and chaired every single meeting. But though the minutes show a variety of points and arguments made by the parties already noted, not a word is heard about Lenin.[34]

Beyond "On Forests," the evidence of Lenin's involvement in other conservation measures is, for the most part, even more fragmentary. "On Conserving Monuments of Nature, Gardens and Parks," another conservation decree signed by Lenin (in September 1921) and cited frequently as one of "his" achievements,[35] was for certain not written by Lenin at all but by members of *Narkompros* (the Peoples' Commissariat of Education)—probably by this organ's subordinate conservation section, the State Committee for the Protection of Monuments of Nature (see Introduction). And the "Forest Code," begun in 1921, well before Lenin's health deteriorated, appears as noted to have been the work of the *Kollegia* (agricultural board) within *Narkomzem*.

It would, of course, be unrealistic to expect Lenin to have personally written the entire text any of these laws. Heads of government do not usually write environmental laws in any country. It is, rather, the claims made on Lenin's behalf that are being questioned here. And to his credit, Lenin does appear to have had some personal input into the drafting and issuance of at least one important document—the 5 April, 1918 circular defending forest specialists (see chapter 5). He also wrote an article dealing with forestry entitled, "On a Unified Economic Plan" ("Ob edinom khoziaistvennom plane"), which touched on themes of planning and efficiency already well explicated elsewhere in this study.[36] Furthermore, certain individuals whose own role in conservation initiatives of the time are not in dispute have given Lenin praise. F. F. Shillinger,

---

[34] GARF. Fond 130, op. 2, del. 1 (4), *SNK Protokoly*. Protokol 107. Zasedaniia Soveta Narodnykh Komissarov ot 2 maia 1918; Ibid., Protokol 111 ot 8 maia; Ibid., Protokol 113 ot 10 maia; Ibid., del. 1 (5), Protokol 114 ot 11 maia; Ibid., Protokol 115 ot 14 maia; Ibid., Protokol 126 ot 28 maia.

[35] This decree is relevant to *zapovedniki*, but not much so to forestry. It has thus not been given any particular attention in this study.

[36] Lenin: *Sochineniia* (Russian), 5th edition, 42: 342.

a game specialist associated with early Soviet hunting legislation, noted in 1932 for example, that Lenin "himself took great interest" in the creation of the 1919 hunting code (also written primarily by N. I. Faleev), which included a ban on the hunting of moose and wild goats and ended the spring and summer open season.[37] Clearly, he had a certain presence within the domain of conservation and resource use.

Far from all forest-related decrees signed by Lenin were pro-conservation, of course. Those directly or indirectly *increasing* forest cutting during the fuel crisis, for example, alone greatly outnumber all those promoting conservation, not just of forests, but in general. Some, noted in chapters 5-6, seem particularly at odds with forest protection. These include: the November 1918 order authorizing "forest sub-sections and land sections" to "carry out unrestrictedly and without compensation the release of forest materials . . . [for] strategic railroad works"[38]; a demand made in May 1919 that the Central Forest Section of *Narkomzem* "give over immediately to the use of *Glavleskom* . . . 250,000 *desiatinas* of forest" situated in a ten-*verst* strip along both sides of railroad lines[39]; and more infamously, the decree "On Timber Procurement"[40] ("O zagotovke lesa"), issued in July 1919 and authorizing the cutting of 375,000 *desiatinas* of forest on erosion-sensitive lands along rivers designated for the transportation of timber by log-floating ("splavnye reka"). And unlike the protection clauses of "On Forests" or other "Leninist" decrees, there is little doubt that these decrees were fully implemented.

Particularly intriguing in regard to questions of Lenin's input to forest organization and conservation laws and initiatives is the almost total absence of direct reference to him by foresters—indeed by anyone—in the forest debates of the day. This is true whether one's sources are official mouthpieces such as *Lesa respubliki* and *Lesopromyshlennoe delo*, academic journals such as *Lesnoi*

---

[37] GARF. Fond A-358, op. 2, del. 7, st. 3.

[38] "Predpisanie SNK lesnym podotdelam i zemel'nym otdelam gubispolkomov proizvodit' besprepiatsvenno i bezvozmezdno otpusk lesnykh materialov Komitetu gosudarstvennykh sooruzhenii i obshchestvennykh rabot dlia proizvodstva dorozhnkyh strategicheskykh rabot," *Dekrety Sovetskoi vlasti*, 4: 91-92.

[39] "Dekret SNK o poriadke otpuska lesa," ibid., 5: 147.

[40] "Dekret SNK ob otpuske Glavnomu lesnomu komitetu 375 tys. desiatin lesosek pod zagotovku lesa," ibid., 5: 338-39.

zhurnal (though this folded in 1918), or the files of the Foresters' Union. The first of these three source categories, of course, frequently vaunts Soviet power; but rarely is there any mention of Lenin himself in connection with forest initiatives. (Regarding "On Forests," Lenin's co-signer Sverdlov himself receives more note.[41]) The archival holdings of the Foresters' Union are especially silent about Lenin. Perhaps the first mention made of him here in any context was in 1920 in the form of an official greeting sent by what was to be the now more compliant foresters' body's last congress, held that spring. The context was the Polish offensive against Bolshevik Russia then underway:

> At this time of a new and burdensome ordeal, at this moment when the Civil War erupts anew thanks to the treacherous attack by Poland, set on to us by the international bourgeoisie, now has gathered the Fourth All-Russian Congress of Forest Workers. The Congress sends you [Lenin] its comradely greetings. The Congress is convinced that the Russian proletariat, inspired by you, will emerge with dignity from this new ordeal and defeat perhaps this last enemy. [Then] will come the time of peaceful construction. Long live worldwide Soviet Power. *Long live Lenin. Down with the Polish lords. All for victory over the protégés of the bourgeoisie* [emphases in original]."[42]

Nor was this a single, token remark. In stark contrast with all earlier Union congresses, this one was marked throughout by the absence of earlier willingness to criticize Bolshevik authority and priorities. Instead, speaker after speaker voted his confidence in the new regime. For the first time also in this forum, use of the word "comrade," previously rare, was ubiquitous. "Comrade Andreeva," for example, read a pro-Bolshevik resolution stating that "We, forest workers, joined together in a union, support . . . without reservation the platform of the Soviet authorities . . . ." The resolution carried on to assure Union members' general agreement "with the measures taken by the Soviet government . . . in connection with the war we are in" and urged members "to apply every strength, effort and [bit of] knowledge" toward fulfilling completely the task of [wood] fuel supply" to help win the war. At the same time, however, this was to be done "with persistent skill so as not to cause irreparable damage to the people's forest economy.[43] Whether the new sentiment was genuine, or simply "protective

---

[41] See, for example, "Lesnaia sektsiia: itogi rabot lesnoi sektsii," 900.

[42] GARF. Fond 5520, op. 4, del. 1, st. 19.

[43] Ibid., st. 5.

coloring,"[44] it was not to be of much use in preserving the Union, which was dissolved soon after.

Indirectly relevant to forest conservation, Lenin's absence is also notable from early accounts of the establishment of the first major Soviet *zapovednik,* in Astrakhan in 1919. The story of how Lenin "took time out" during the depths of the Civil War that January to meet with N. N. Pod"iapol'skii (a local Soviet official from Astrakhan) to hear his "case for conservation"—and to examine the blueprint for this pioneering Soviet *zapovednik* he had brought—is perhaps the single best-known of all the Lenin-as-conservationist stories, and has been repeated by western authors also.[45] Lenin supposedly played a key role in then helping turn Pod"iapol'skii's plans into reality.[46] Thus it comes as something of a surprise to find no mention of the meeting, or any reference to Lenin at all, in any of the several accounts of the *zapovednik*'s creation written over the few years immediately following 1919.[47] (Possibly the first reference to either is by

---

[44] Douglas Weiner coined this term to describe the adoption by conservationists of "politically correct" language and so on, intended to allow them to exist relatively unmolested in the increasingly stifling intellectual environment of Stalin's Cultural Revolution. I have used the term more loosely here.

[45] See, for example, Douglas R. Weiner, *Models of Nature* ("A Fateful Meeting," in chapter 2).

[46] V. E. Boreiko has recently challenged some of the details, though not the basic facts, of the standard account. See his "Ekologichskii Leninizm." In *Belye piatna,* 133.

[47] Accounts of the establishment of the Astrakhan *zapovednik* I have seen are:

1). A long essay written in 1920 by R. Pavlinov (GARF A-358, op. 2, del. 2, st.st. 5ff). This makes no mention of Lenin but does note that Pod"iapol'skii's *zapovednik* plans received a favorable hearing f rom the Museum section of the Education commissariat, the organ then primarily responsible for conservation.

2). A shorter, untitled, note of 1920 immediately following in the same source.

3). An article on Astrakhan *zapovedniki* published in 1924 in *Kommunist, organ Astrakhanskogo gubkoma . . . ,* 165 (25 June 1924), 1752,. I have seen only the version of this deposited in GARF. Fond A-358, op. 2, del. 2. following the above two sources.

4). An article by A. Diunin entitled, "Astrakhanskii gosudarstvennyi zapovednik" (copy in GARF A-358, op. 2, del. 2. st.st. 28-29), taken from journal *Nash krai*" 1 (1927), 2. (Astrakhan'. Tipo-lit.).

Pod"iapol'skii himself in 1929.[48]) Particularly intriguing is that Lenin is not mentioned even in a report on this *zapovednik*'s creation written by Pod"iapol'skii himself only days after the supposed meeting.[49] Although none of these circumstances prove that the story of Lenin and Pod"iapol'skii is apocryphal they do raise suspicions of after-the-fact embellishment, at least. Why, one wonders, was Lenin's role not worth noting at all in 1919, worthy only of a brief mention in a single journal article ten years later, but of a starring role in a whole series of books and essays decades later?

If Lenin's contributions to early Soviet conservation remain unclear, a probable rough picture can be inferred in which Lenin did indeed make significant contributions, though ones far less spectacular and a lot more indirect than is usually asserted. First, and undoubtedly, forest management and conservation during the period 1917-21 were conceived within the general Leninist parameters of resource nationalization and centralized planning along scientific-technical lines. Forest reform in the context of 1923's "Forest Code" fit in with the altered—but equally Leninist—landscape of the NEP. Lenin, then, as the leader of the Bolsheviks and primary architect of early Soviet policy, could not but be also a shaper of conservation policy. Beyond this, as already noted, he would almost certainly have exercised at least some influence on relevant legislation passed by the SNK and STO, though again, precisely how much remains far from clear. Beyond this, however, his contribution has probably been considerably exaggerated.

Since Lenin probably set some of the tone for early Soviet forest conservation, it is useful to ask also what *kind* of conservationist he was. Despite the efforts of biographers to create him at least in part as a sentimentalist, driven by feelings of love for the beauty and power of wild nature—and even by affection for his fellow creatures—his actions mark him out as a clear socialist "utilitarian." In other words, Lenin was a man for whom nature was first and foremost a storehouse of raw materials, the value of which could be realized primarily by transforming them (by human labor, and in accordance with Marx's

---

[48] N. N. Pod"iapol'skii, "Vladimir Il'ich i okhrana priroda," *Okhrana prirody* 2 (1929): 35-38.

[49] See, "V nauchnyi otdel narodnogo kommissariata prosveshcheniia. Chlenogo Agronoma N. N. Pod"iapol'skogo. Ob ustroistve zapovednikov v Astrokhanskom krae. Doklad." GARF. Fond A-358, op. 2, del. 2 st.st. 2-3. This note is dated 25 January. The meeting with Lenin is supposed to have taken place on 16 January.

labor theory of value) into the material underpinnings of a socialist economy, that is, into appropriate industrial and military infrastructure and manufactures. These priorities are clearly evident as the basic undergirding of early Soviet forest organization in general—for example, in the aforementioned stated goal of drawing from the forests 100% of the annual rate of growth for supplying the rising demands of state and domestic users and for export. This will be discussed further in the concluding essay, however. Here, we are more concerned with Lenin's *personal* primary interest in forests as sources of raw material for socialist construction. This can be glimpsed frequently: in his 1921 comment that "from the northern forests it would be possible in the very near term to get half a billion dollars [*polmilliard valiuty*]" on the export market[50]; in a conversation with Gorky in London in 1907 in which Lenin used wasteful timber cutting practices—especially the profusion of chips, small branches, and stumps left at the scene of cutting—as his primary example of the "anarchy of production" under capitalism, which the efficiency of socialist production would eventually grossly out-compete (this is told in numerous secondary Soviet accounts); in his readiness to solve the fuel crisis at the expense of massive over-cutting (see chapter 5); and so on.

Lenin had similarly utilitarian priorities regarding other natural resources, too. According to a hunting companion from his days of exile in the late 1890s, "Vladimir Il'ich . . . like no one else . . . fully understood the economic significance for the proletarian state of a correctly-organized hunting economy"[51] (this may also be after-the-fact myth-making, of course). Instructive also is a decree Lenin signed on 22 February, 1919 entitled "On Forbidding the Slaughter of Horses for Meat." (*O zapreshchenii uboia loshadei na miaso*). Echoing official anger at the time with the onslaught against forests, this decree condemned the ongoing "massive and lawless slaughter of horses for meat" by local populations, as well as deliberate "laming or harming" of fit horses in order to justify their slaughter for the same purpose. Needs of state, not humane concerns, were at issue, however. The loss of serviceable animals "threatened with catastrophe the armies' supply of horses and [endangered also] the agricultural economy which,

[50] Quoted in RSFSR, *XVI S'ezd vsesoiuznoi kommunisticheskoi partii (b) 26 iiunia - 13 iiulia 1930, Stenograficheskii otchet*. Partizdat TsK VKP(b), 1935: 309.

[51] "V. I. Lenin kak okhotnik," *Okhotnik: Illustrirovannyi zhurnal okhoty i rybolovstva* 10 (1927): 31.

come spring, will not have sufficient horse power for cultivation of the soil."[52] In May, at a time of need, Lenin did not hesitate to "mobilize right away [and] everywhere possible . . . all horses [serving] both artillery and cavalry . . . leaving only an absolutely necessary [minimum] quantity" for draft and other purposes.[53] Not once did the man who had earlier refused to shoot a fox, who had never harmed birds' nests (and so on), ever utter a single word urging the humane treatment of animals or expressing any regret at having to use them so harshly. Nor should we be surprised by this. Lenin the realistic and hard-headed leader of men and revolutions was always a more true-to-life historical portrait than Lenin the animal-lover or wilderness enthusiast.

It can easily be argued that Lenin saw all the Republic's natural resources—and certainly forests—in the same way he saw horses, as the necessary tools with which to carry out the vital military, economic, and industrial tasks of the day. Indeed, a very great number, probably a significant majority, of all decrees signed by Lenin related either to forest or wildlife conservation, began by announcing identical, and similarly utilitarian, aims: to increase the amount—and improve the quality—of resources available to the Soviet state. It is apparent, in fact, that in matters of conservation and exploitation, Lenin personally—and early Soviet policy in general—made little if any distinction between domesticated and wild nature. All alike—from cows to forests—were simply resources to be commandeered, used, and planned for the maximum material benefit of the state. Thus, numerous orders issued to the population to turn domestic animals or crops over to the state (under the policy of War Communism) mirror in their basic language and stated goals decrees such as "On Forests" and the "Forest Code" claiming all of wild nature for the state and proposing to meet all necessary material demands. None of this, of course, makes Lenin an exceptional environmental rogue. American and other western politicians and industrialists of the time and later evinced similar priorities.

---

[52] "Dekret SNK o zapreshchenii uboiia loshadei na miaso," *Dekrety sovetskoi vlasti*, 4: 200.

[53] "Postanovlenie Soveta Oborony o srochnoi mobilizatsii loshadei," ibid., 5: 90.

IX
Epilogue:
Forest Conservation Between Lenin and Stalin, 1925-1929

Lenin's death in 1924 had no immediate or obvious effect on the direction of forest management or conservation. However, while the previous chapter has argued that Lenin's vision and involvement in forest affairs were quite limited, they certainly surpassed those obtaining among the Soviet leadership over the following half-decade. The key players in Moscow manoeuvring for power—Kamenev, Zinoviev, Trotsky, and especially Stalin—made no obvious contributions to forestry during the 1920s. There were to be no further major legislative initiatives, no notable efforts to implement existing measures, nor even any rousing directives to the forest community such as "Lenin's" decree of 5 April (see chapter 5). From 1925, once the fallout from the new "Forest Code" had largely settled, through to the unleashing of a brutal whirlwind of forces in the countryside during Collectivization (beginning in earnest in 1929), the period—in forestry at least—is best characterized as one of drift. Perhaps predictably, the problems that had hamstrung Bolshevik implementation of rational, national forest use and conservation—even at the reduced levels prescribed in the "Forest Code"—persisted with remarkably little alteration. Official figures released in 1927 showing annual rates of unauthorized forest cutting, provide just one of many possible examples: 3,657,000 cubic meters in 1923-1924; 3,501,000 in 1924-1925; 2,870,000 in 1925-1926; and 3,475,000 in 1926-1927.[1] The laying-off of large numbers of forest guards in 1926-27 may have contributed to the reversal for 1926-1927 of modest improvements made in earlier years.[2]

This study has cited a variety of reasons to explain the state's failure to make substantial progress—bureaucratic inefficiencies, insufficient resources, the fuel crisis, and so on—but the primary cause, it should be clear, was and remained the resistance of rural populations. And during the middle 1920s the issue appears to have been allowed to slide. As late as 1929, a review of the impact of the "Forest Code" in the Soviet Central Executive Committee had

---

[1] A. Denis'ev, "O bor'be s samovol'nymi porubkami lesa," *Lesnoe khoziaistvo i lesnaia promyshlennost'* 1 (1928), 53.

[2] Ibid., 54.

only the most familiar tales to tell: of general disorganization, peasant dissatisfaction, and ongoing unauthorized cutting. Noted one speaker:

> During 1925-27 forests of local significance were given over as the property of the peasantry. And what happened? In some places we gave [forests] and in some places we didn't and in so doing created among the peasantry a certain dissatisfaction. (Voice: That's true!) Certain peasants have forest and sell [timber] on the market. Peasants from other villages look bitterly at those who have forests and go to the forests of [general] state significance, steal timber from there and sell it on the market too. (Voice: That's true!).[3]

This was followed by what had become the almost hopeless mantras of the forest authorities: calls to educate local populations, to "correct" relevant "errors," and "not [to] create dissatisfaction among the peasants."[4] Significantly, however, these may have been among the last such statements. For by now a fundamental shift in state-society politics on a much wider scale than just forestry was already underway. Stalinization—and in particular the murderous program of collectivizing agriculture which was one of its central elements—would by about 1933 result in the almost total destruction of the peasantry as a formidable, independent, and hostile political culture.[5] Questions of their appeasement would never again have the same resonance; and the basic nature of forestry would change significantly. These issues are the stuff of a separate study and will be dealt with only in outline here.

A brief survey of the second half of the 1920s suggests strongly that the "Forest Code" had a less-than-transforming effect on the efficiency and organization also of *authorized* forest cutting by state organs. Little real progress, for example, seems to have been made after 1923 toward the Code's

---

[3] RSFSR. *Vserossiiskii tsentral'nyi ispolnitel'nyi komitet,* XIV sozyv. Vtoraia sessiia.. Zasedanie deviatoe. *Biulletin'* 9 (24 November, 1929), 9.

[4] Ibid.

[5] During 1929-1933 in particular, the Soviet government compelled the peasantry to give up its small individual or locally-organized collective farms (including animals, crops, and equipment) and join large-scale collective or state farms. The process was rapid and brutal. During these four years alone, perhaps as many as 5 million persons (mostly "wealthy" peasants, or *kulaks*) were deported or disappeared. Since peasants preferred in many cases to slaughter rather than surrender their livestock, the animal population declined spectacularly too. Nicholas V. Riasanovsky reckoned that from 1929-1933 the horse population of the Soviet Union fell from 34 million to 16.6 million, while sheep and goats were reduced from over 147 million to under 51 million *(A History of Russia,* 4th edition [Oxford University Press, 1984], 551). The result was a total transformation of the social composition of Russia, with the peasants effectively proletarianized and their resistance shattered.

ultimate goal of annually harvesting 100% of national timber growth. The figure had been slightly over 20% in 1924.[6] It was reckoned unchanged in 1929.[7] (Unfortunately, figures are not available for the pre-Revolutionary period, or for the era of the Revolution and Civil War.) Ironically (given Soviet goals of nationally-planned forestry), since total yearly timber growth far exceeded the total amount harvested, there was, in fact, no over-cutting at all on a national level; and there existed more than sufficient resources to meet all demand. On the other hand, with scant progress being made in opening up new resources (alike in the remote regions of the north or east and in less accessible areas in European Russia, away from rivers, railroads, and so on), and with the loss of large areas of forest to local use and a concomitant failure to establish rational use systems around the countryside, almost the entire harvest continued to be drawn from relatively small and already heavily-depleted forests along railroads and rivers and near villages, towns, and cities. A 1929 report noted in this regard that the "rapid development of forest exploitation, notable growth of procurement and export of timber" achieved since about 1921 had been possible only because "the extraction of timber [in exploited areas] exceeds the natural yearly growth . . . . That is, we are working up a deficit . . . we are over-cutting the forests."[8] Funding shortfalls, bureaucratic inefficiencies, and a shortage of skilled technical personnel all appear to have remained unresolved as well, and were the objects of persistent

---

[6] A 100% cut of all Soviet forests (not just Russian), it was estimated in 1924, would equal some 50,609,000 "kuby" (see chapter 6, n. 29). More than half of this area was considered inaccessible [*neudobnyi*], however, leaving a much lower potentially exploitable total of 24,058,000 *kuby* (47.6% of total estimated annual growth). Organizational, technical, and logistic constraints further reduced actual targets that year to 11 million *kuby*. This was the largest target figure to date, however, and represented approximately 21-22% of total annual growth, or about 45% of accessible growth. See RSFSR. *Vserossiiskii tsentral'nyi ispolnitel'nyi komitet,* XI sozyv. Tret'ia sessiia. Zasedanie vos'moe (15 October, 1924), 348.

[7] RSFSR. *Vserossiiskii tsentral'nyi ispolnitel'nyi komitet,* XIV sozyv. Vtoraia sessiia. Zasedanie deviatoe. *Biulletin'* 9. (24 November, 1929), 5. Absolute timber harvests, on the other hand, did grow during this period. The rise in absolute numbers without a corresponding rise in the percentage of new growth cut is explained in at least two ways. First, the total area of forest under Soviet control grew during the 1920s as the state expanded its borders. Second, and perhaps more important, the practice of "borrowing" cuts from future years allowed greater numbers of trees to be harvested without it representing a greater percentage of total growth in any one year.

[8] RSFSR. *Vserossiiskii tsentral'nyi ispolnitel'nyi komitet,* XIV sozyv. Vtoraia sessiia. Zasedanie vos'moe. *Biulletin'* 8 (24 November, 1929), 7.

recriminations and regret throughout the 1920s.

It will be useful here to say something also about developments in the few forest *zapovedniki*—those specially-designated areas of land enjoying, at least officially, extraordinary protected status and which were outlined in the introductory chapter. Overall, *zapovedniki* appear to have suffered similar problems to forests in general. Particularly compelling is the following assessment written in 1926 by N. A. Kuznetsov and published in the journal *Priroda*:

> Of the protection of monuments of nature [and] the creation of *zapovedniki*, much has been written by us since even before the Revolution; there have also been created no small number of organs for the protection of these 'monuments of nature' [see chapter 1]. The Revolution created new organs for the protection of nature and created its own literature on nature conservation. But at the same time that in America there have been created "national parks" in which conservation measures are carried out on a typically American broad scale, at the same time that in Germany, regardless of war and revolution, the Germans continue to protect particular trees and rock formations [*valuny*] with typical German pedantry [*pedantichnost'*], here in Russia the business of nature protection before and now has been coming along Russian-style, that is, we write, establish *zapovedniki*, speak about them at numerous conferences, organize various nature-preservation committees and commissions, while the protected "monuments of nature," completely independent of these learned discussions and legislative acts, continue to live existences very different from the ideal of "monuments of nature."

To illustrate his point Kuznetsov turned specifically to forests, and to the Vorskla Forest *zapovednik* established in Kursk province in 1922. Vorskla was, he said, a typical example of *zapovedniki* in general, including in its failure "almost completely" to meet even "the most basic elements of the concept of *zapovedniki*."

Do many [Russian] naturalists know of the existence of this forest *zapovednik* . . . ? Very few, it seems. And is there being carried out there annual, systematic research of its soil cover, vegetation, and animal population[s]? Evidently not, for there are neither the resources nor sufficient numbers of experienced research-specialists. As far as the actual preservation of nature in this *zapovednik* [goes], that is set up rather primitively and does not correspond to the basic significance of the *zapovednik*. The forest has been declared protected, but to this day no buffer strip has been marked out to define the borders of this area. . . . and the local peasant *mir* population continues its insolent [*naglyi*] cutting and plunder of the forest and its cattle grazing.

Local organs, again, were part of the problem: "In 1924, at the order of the local authorities 120 ancient [*vekovye*] oaks were cut down right in the most protected area [of the reserve], and despite energetic protest by the ecological station there, a belt of alder trees protecting the southern edge of the forest from hot, dry winds was also felled."[9] Kuznetsov finished his remarks by calling once again for nature protection "not only on paper, but in fact."[10] A 1924 report on conditions in the Crimean State Forest *zapovednik* similarly notes the continuation of cutting even within its most strictly protected areas.[11] The pattern, indeed, recurs almost wherever one looks.

The story of Soviet forestry does not end, of course, in the late 1920s. However, the collectivization of the peasantry that began at this time so fundamentally recast the basic social landscape within which early Soviet forestry had existed as to make it clearly a watershed and a convenient terminus point for the present account. Nonetheless, there was to be much continuity after 1929 also. While forest exploitation retained and even increased the high priority inevitable in any industrial or industrializing state, efforts continued in particular to establish and implement planning and use on a national scale, and within this framework to conserve forests.

The year 1929 of course brought the full force of the First Five-Year Plan to Soviet industry and agriculture. In forestry as elsewhere this entailed many specific changes that go far beyond the scope of this book. Overall, however, it is clear that the Plan brought an increased emphasis to extraction rather to conservation, and also to the mechanization of the former process. Though neither conservation talk nor action were extinguished, the most productive years, and the greatest opportunities were already in the past. In 1929 the SNK decreed an end to the system that had prevailed in law before, whereby timber procurements were worked out on a year-by-year basis and instead granted long term rights to extract timber from the country's central regions to the chief economic organ (the VSNKh) and to the People's Commissariat of Transport (*Narodnyi Kommissariat Putei Soobshcheniia*, or NKPS). This provided cover for further over-cutting, among other results. The ascendancy of the VSNKh and of extractive and exploitative interests in general over the forest apparatus of *Narkomzem* and its mission of administration and conservation was to accelerate during 1929-1931. The second session of the Soviet Central Executive Committee held in November 1929, for example, resolved to centralize all forestry functions under the VSNKh. This was decreed officially *by the SNK RSFSR* on 25 February, 1930. Even so-called "experimental forests" (*opytnye lesa*—areas given over for

---

[9] N. [A.] Kuznetsov, "Ves'ma vazhnyi vopros pamiatnik prirody," 98-99.

[10] Ibid.

[11] Professor Sukachev, "Krymskii Gosudarstvennyi Lesnoi Zapovednik," *Lesovod* 2-3 (1924), 29.

scientific research, often within *zapovedniki*) were to be handed over[12] as the forest department of *Narkomzem* was pushed increasingly into a largely "consultative" and clearly subordinate role. The "Forest Code" itself was essentially to be abandoned in 1931 as part of a further reorganization that divided the forest fund into either "forest-industrial" (*lesopromyshlennye*) or "forest cultural" (*lesokul'turnye*) categories[13]—a development that also was designed largely to facilitate further exploitation.

[12] "Vse lesnoe khoziaistvo—v VSNKh," *Lesnoi spetsialist* 5 (1930), 4.

[13] *Lesnoe khoziaistvo SSSR za 50 let, 1917*-1967, 19.

X
Conclusions

Early Soviet forest conservation presents a variety of general features. First, it was far from an entirely new creation, born out of the October Revolution. Rather, it derived significantly from models and programs worked out earlier, especially during the 1900s and 1910s. Examples abound: the idea of splitting forests into protected and unprotected categories; of increasing the role of the government and ultimately of transferring ownership entirely to the state; of better satisfying local populations' needs and concerns; and in particular, of establishing all the forests as one "national inheritance" to be administered according to a single, scientific plan. All these concepts were current and championed throughout the decade or so before October 1917 by a great many foresters and other interested parties. (Early Soviet forest law even retained or regurgitated numerous articles from tsarist antecedents such as the Statute of 1888.[1]) Thus, the Bolsheviks' main role was not to establish basic new concepts or parameters—these they largely inherited—but to choose within existing ones and hone a general "foresters' consensus" into a clearer, more narrowly-defined vision which could then hopefully be implemented. As already argued, in this connection the Bolsheviks' chief contribution was to resolve in favor of the center (at the expense of local organs and populations) existing ambiguities regarding in whom state ownership and control should be vested. What the Bolsheviks *do* appear to have originated, however, was the fateful subordination of specialists to politicians discussed in chapters 4 and 5 (and perhaps also the integration of exploitation and protection into a single plan and bureaucratic structure, as outlined in chapter 6—though in the absence of further research this is less clear). Viewed in this light, however, February—not October—stands as the critical break in Russian forest history: February cleared away the old social, legal, and political edifice, making substantial change possible. And as argued in

[1] An article published in 1923, for example, noted numerous clauses shared more or less in common by the "Forest Code" and the 1888 "Statute on the Safeguarding of Forests." See P. Serebrennikov, "O novom lesnom kodekse," 10-11.

chapter 2, under the Provisional Government between February and October 1917 (old style), considerable efforts and some progress were made toward achieving some of the major forest goals articulated earlier. February, of course, also unleashed the ruinous centrifugal forces the Bolsheviks were to struggle against and compromise with until the advent of Collectivization.

These latter forces—overwhelmingly in the form of peasant non-cooperation or even active resistance—constitute the most obvious and also the most consequential feature of Lenin-era forest organization and conservation. Because of them, the Bolsheviks made relatively little headway—except on paper—in most of their relevant endeavors. In forest protection as in so many aspects of early Soviet life, the basic reality was that government and local populations—state and society—were pursuing very different, perhaps irreconcilable, interests. Efforts to implement rational, national forest management and scientific conservation flew in the face of popular culture and of local aspirations to use the forests more or less without restriction or plan. Likewise, politicians' and educated foresters' notions of trees as scarce resources in need of careful husbanding and prudent planning clashed endlessly with peasant conceptions of them as inexhaustible "gifts of God" restored to their rightful peasant proprietors by the "Tsar-Liberator," Alexander II at the time of the Emancipation in 1861. Prior to the end of the 1920s, the government lacked the power effectively to alter this dynamic, forcing it into an increasing compromise represented best of all by the massive transfers of forest—from general-state to local peasant holdings—carried out under the auspices of the 1923 "Forest Code." By abandoning as unworkable for the time being the ultimate goal of a single centrally-planned (but locally flexible) forest economy, Bolshevik authorities were in many respects reluctantly accepting de facto decentralized forest control as a *fait accompli*, and recognizing the futility, at least in the early 1920s, of trying to change this reality.

Elements of this kind of compromise were not completely absent even from the earlier, more heavy-handed and centralizing "On Forests." Here too, exceptions had been written into the law allowing certain "peasant agricultural secondary forests" to be "excluded [from the action of the law] as 'non-forests,'" and subordinated instead to the "Basic Law on the Socialization of Land."[2] The scale of compromise, however, differed considerably. "On Forests" gave over to local control only certain "trivial planted areas [*melkye nasazhdeniia*], not

[2] N. I. Faleev, T. Nekhoroshev, D. Sazhin, "Lesnye organy," 332.

deserving of the name 'forest,' on parcels of peasant land."[3] The "Forest Code," on the other hand, transferred as many as 19 million hectares to local use.

A general cultural clash between the state and rural populations was not the only source of tension, however. Pragmatic realities set the two constituencies against each other. For example, Comrade Chudinov—a representative of a rural Soviet, speaking to the Party Central Committee in 1928—quoted a peasant he had recently met: "You [the state] think to take the forest in order to carry out proper [*pravil'noe*] forestry, but from this 'proper' forestry we get no benefit."[4] That this was probably true was not lost on the Bolsheviks, as has been seen. In fact, by the early 1920s and for several years thereafter, state and peasant appear to have been in considerable agreement that what most peasants had actually got from early Soviet forest planning and conservation were the manifold inconveniences associated either with having to travel to designated cutting areas or timber stores to get (or to not get) exactly the same supplies they could have got locally and spontaneously, or in other ways having to deal with this system, and with its bureaucracy and personnel. While for rural populations this might have been sufficient evidence that Soviet efforts to rationally and centrally manage Russia's forests should be abandoned, for the Soviet authorities and professional foresters alike it was evidence that forest planning needed to be better organized and more properly enforced. Moreover, for the central political authorities at least—and probably for the forester community as well—allowing

---

[3] Ibid., 334. I have not come across any figures for the amount of land thus designated or transferred. It appears to have been small. Certainly, it does not compare with the amount transferred after 1923. That this type of division had not been embraced on a wider scale until 1923 was in large measure due to the desire of the center, of Faleev, and of certain other interested parties, not to concede to peasant localism all hope—or the entire reality—of a unified forest economy (and thus a single system of protection.) Faleev, for one, clearly hoped that by including both local, agricultural woodlands and Russia's larger forest massifs under one legal roof—in tandem with appropriate forest propaganda—"On Forests" would ultimately help wean the population off its localisms and motivate it to embrace a more truly "socialist" view of the forest economy and their place in it. See, for example, Faleev's comments in ibid., 333.

[4] RSFSR. *Vserossiiskii tsentral'nyi ispolnitel'nyi komitet*. XIII sozyv. Vtororia sessiia. Zasedanie desiatoe (4 April, 1928), 469.

relatively unfettered local use of forests was always going to be inadmissible. For the former this would have meant setting the forest world aside from the goals and methods of socialist construction being applied more generally in the new state—in effect creating a "special zone" whose economic, social, and political realities did not conform with those in other domains. For foresters it would not only have gone against everything they believed in and stood for as technical experts, but also would have meant that they and their organs had no real purpose any more. In both cases simple self-interest set rural populations and the central authorities against each other.

One of the ironies of the whole situation is that there were within Soviet borders more than ample forest resources to meet all present needs as well as those for many years to come. State and society could both have been served, at least in theory. Matters were hopelessly complicated, however, by the fact that forests were most abundant where people were least so, and vice-versa. Success in matching supply and demand, therefore, could only be realized by implementing exactly the type of planned, national system "On Forests" envisioned. Yet this was precisely what local and particular interests—real or perceived—so completely rejected. The huge difficulties inherent in trying to inventory, access, and transport forest resources in the far north and Siberia compounded the difficulties of implementing rational forestry, as did also the under-development of the Russian economy in general.

Nonetheless, if implementation of forest planning or conservation was problematical, lax, and unsuccessful overall, it was not entirely so. One piece of evidence of this is the frequency of complaints by peasants sent what they felt were unreasonable distances to collect or cut wood, or regarding the onerous conditions of use thereafter, and so on. Had the peasants been entirely immune from actual regulation and—as they saw it—interference by the state, such complaints would surely not have existed; or at least they would have been much reduced. Conservation laws *did* function and *were* in many cases enforced, then. But they were also ignored or violated on a pandemic scale.

Some evidence of progress can be seen also in official forest-related statistics. Of particular note, forest planting, which from 1844 to 1917 had been carried out on an average of 16,000 hectares per year, rose over the period 1917-1927 to an annual average of 43,500 according to official Soviet figures.[5] There was, again following official figures, also a more modest rise over a similar

[5] *Lesnoe khoziaistvo SSSR za 50 let, 1917-1967*, 103.

period in the quantity of land reclaimed from swamp. Over following decades Soviet figures would come to surpass—often by large margins—tsarist ones in a variety of categories: area of forests officially protected, numbers of trained specialists and qualified personnel, resources devoted to reducing fire and insect-infestation damage, the quantity of forest-conservation legislative acts passed, and so on. In 1929, however, most of these achievements still lay far in the future. And given the later growth of the country's technological base, economy, and population, a mere juxtaposition of nineteenth- and twentieth-century numbers in this manner is of little real value.

Even if early Soviet conservation laws *had* been fully implemented and properly enforced, however, it is likely the results would still have disappointed. For the laws themselves were deeply flawed. For example, in comparison with conservation regimes in other times and places, Soviet ones—in forestry at least—offered remarkably little separation between conservation and exploitation. "On Forests," for example, assigned the creation and administration of planning for forest exploitation *and* conservation alike to the central forest apparatus of *Narkomzem*. Unlike in the western democracies (and especially in more recent times) there was here no mechanism by which forest conservationists could effectively challenge or check forest users. This was, of course, not an oversight but a definite intention. Lenin himself characterized the Soviet state's fundamental approach to nature protection as the latter's "organization within the very process of the exploitation of natural resources, that is, the rational use thereof based on thoroughly-worked out scientific-technical rules."[6] A writer for *Lesa respubliki* made a similar point in 1918, noting in a discussion of "On Forests," that it was

> time to renounce the thought that the task of forest legislation is forest protection and only forest protection. Even for the most superficial reader it is clear that the Basic Law [on Forests] has moved away from police guardianship [which had been the case with 1888 Statute, for example], from regulation, from a system of bans, [and] that it has entered onto a path of *organic creative work* [emphasis added] . . . ."[7]

What was envisioned then—and legally and organizationally implemented—was not a system of checks and balances, with exploitation activities and organs being held accountable to policing by independent

[6] P. I. Zhukov, *Leninskye printsipy okhrany prirody*, 7.

[7] *Lesa respubliki* 10-11 (1918), 548.

conservation laws and organs, but a seamless web of production, exploitation, and protection. If this makes some sense in theory, in practice it defanged conservation which, in the absence of a separation of powers, and in an environment placing great emphasis on economic growth, lost out handily to exploitation. The lack of a mechanism for calling to account the destructive activities of the *zheleskomy* provides one among many examples of this. Thus, though conservation organs did exist (see chapter 1) these had little authority. Moreover, they functioned (fairly impotently) mostly in the world of *zapovedniki*. In the largely separate and certainly much broader realm of forestry they were virtually non-existent.

Bolshevik forest conservation legislation was still further weakened by the fact that the very vehicles through which conservation might have been usefully established—protected forests, for example—themselves lacked any firm foundation in law. "Protected forests," for example, existed according to "On Forests" only as long as deemed "necessary" (article 86) Who decided on matters of "necessity"? Not conservation organs but the general Soviet leadership—whose decision in such matters was not subject to review or challenge by conservationists, and whose priorities lay with industrialization and electrification. Similar conditions hamstrung conservation clauses in the "Forest Code" also. Indeed, the entire corpus of forest legislation, even regarding *zapovedniki*, was similarly afflicted. Thus, regardless of the practical realities undermining work in the field, *even on paper* early Soviet forest protection lacked bite. The split of forest affairs into *lesokhoziaistvo* and *lesopromyshlennost'*—discussed in chapters 5 and 6—though perhaps appearing at first glance to represent some kind of division of protection from exploitation, in fact had only a little of this about it. Certainly, *lesopromyshlennost'* involved mostly exploitation. But *lesokhoziaistvo*—represented nowhere better than in the two major forest law codes of 1918 and 1923—encompassed both conservation and exploitation conceived as a single whole. In fact, it is better to understand *lesokhoziaistvo* as the domain of theory and planning, and *lesopromyshlennost'* (including the activities of the *zheleskomy* and other economic organs) as "practice." The latter, of course, was clearly victorious whenever there was any conflict, such as during the fuel crisis of 1918-1919 (see chapter 6).

Early Soviet forest conservation can be better understood also when viewed comparatively—for example, against analagous regimes in other European states and in North America. Though this properly requires a separate study, a few brief remarks can be made here. I have argued that the central problem of early Soviet forestry was the great disunity of vision and interest separating central authorities from rural populations. Even the most casual glance at nineteenth- or twentieth-century forest affairs in America, Britain, France, Germany, Scandinavia, Switzerland, and so on, reveals the complete uniqueness of the Russian situation in this regard. If individuals and private or state organs in all of these other countries (especially the last four) had by the early twentieth century put forward or realized at least some level of forest conservation, none came close to the Soviet vision in terms of sheer ambition. None, for example, was as truly national, and certainly none (even in Germany) promoted the need for science and planning to a comparable degree. At the same time, and critically, none of these countries had a peasant population so predominant as Russia's, or anywhere near so thoroughly bound—socially, psychologically, and economically (if no longer legally)—to the soil and often to the forests themselves. The combined effects of centuries of land enclosure (especially in Britain during the fifteenth to nineteenth centuries), of agricultural and industrial revolutions, urbanization, and so on across much of Europe—all had long since largely broken the relationship of most European peasants to the land (albeit it to widely varying degrees) and brought them at least somewhat into "modernity." This was not so in Russia. In other words, Russia's central forest apparatus's plans were unprecedentedly progressive while its population was singularly conservative. Nowhere can one find a second instance wherein aspirations and realities were so hugely misaligned.[8] Viewed in this light, the relative lack of concessions to

[8] Not all European states were equally different from Russia, of course. Perhaps the closest situation to Russia's in this regard is that of France. In 1669 Louis XIV's chief minister Jean Baptiste Colbert passed perhaps the first comprehensive forest code of any country. It established an administration for the entire state and laid out rules or parameters for numerous forest-related practices, including a ban on all unauthorized forest clearing. In its desire to plan and regulate forest activity, however, the law, not surprisingly, was considerably less ambitious than "On Forests" or even the "Forest Code" and is more usefully compared to the 1888 "Statute" passed in Russia under tsar Alexander III. See, for example, Varenne de Fenille, P. C. M., *Mémoires sur l'administration forestière* (A Bourg, De l'impr. de C. C. G Philipon, 1792); or, Hubert Babeau, *La Forête de Chizé* (Poitiers: Étude geographique. Université de Poitiers, 1973). Other, relatively well- or ambitiously-managed forests prior to the early twentieth century include the city forest of Zurich, Switzerland, which is the world's oldest continuously-managed one. Scandinavian and German forests also have considerable, though somewhat less venerable traditions of management. In none of these areas, even eastern parts of Germany, were peasant populations or forest-organizational ambitions anywhere near as ill-matched as in Russia. Britain, which had become mostly

peasant and localist sentiment in "On Forests" and even in the "Forest Code"—or put another way, the early Soviet emphasis on the national good and the central plan over individual interests—whether construed as courageous fidelity to a lofty ideal, or as blind arrogance, was its most abiding weakness.

All this raises two questions with major implications outside the immediate domain of early Soviet forest conservation. First, regarding the transition from Leninist to Stalinist Russia in general, might the massive disconnect between the nascent Bolshevik state and rural society on matters of forestry have in any way helped inspire the violent assault launched by the former against the latter in the form of Collectivization beginning in 1928? In other words, should forest-related tensions be added to the usual list of factors cited by historians who seek to

---

deforested long before forest conservation was even dreamed of there, is in this regard among the most backward of Europe's nations. While certain local forests had been protected going back as far as the Norman Conquest, the issue was taken seriously at a national level only after the First World War, when German submarines' successes at preventing timber imports underscored the need to secure domestic supplies by means of strict conservation and major reforestation. Little was done, however, until 1945, when Parliament passed legislation envisioning the protection or creation of 5 million acres of forest over the following fifty years. A somewhat different picture obtains for the British Empire, however, which by the turn of the twentieth century had more protected territory (not all forest, of course) than any other political entity in the world.

explain the state's decision in favor of Collectivization?[9] A proper answer to this question would require clear and direct evidence, which is currently lacking. Circumstantial factors, however suggest that the question may be worth pursuing. With the failures of both the War Communistic approach enshrined in "On Forests" and of the more conciliatory NEP tack of the "Forest Code" one can certainly well imagine the appeal to some within the Bolshevik forest apparatus of much more decisive measures. At the very least, the profound tensions evident in forest politics between state and society clearly present a useful case-study of the wider disconnect in early Soviet Russia whereby a minority of urban socialists, committed to ending private property and to marshalling the energies of individuals for collective goals, held political power in a state whose vast majority were peasants committed to retaining and expanding private property (albeit frequently in the form of village-based collectives rather than individual holdings), and who were also committed to local and small-group, instead of

---

[9] Several authors have explained Collectivization in economic terms—as a way of improving the efficiency and output of the rural economy in order to finance rapid industrialization. See, for example, Roy D. and Betty A. Laird, *Soviet Communism and Agrarian Revolution* (Harmondsworth, UK: Penguin, 1970). Others, such as Moshe Lewin, have presented Collectivization as a matter more of necessity more than choice, pushed on the Party in general and Stalin in particular by the realities of a "grain crisis" in 1928-1929, whereby the countryside provided too little food for the cities. See, Moshe Lewin, *Russian Peasants and Soviet Power: A Study of Collectivization* (Evanston, IL: Northwestern University Press, 1968), 250-51.) Robert Conquest, on the other hand, has largely dismissed the "grain crisis" and any notion it forced the Bolsheviks' hand against the peasantry. Rather he sees in 1927 and 1928 only a "temporary disequilibrium in the grain market, easily correctable if normal measures had been applied." Bolshevik paranoia, fear of the market, ignorance of real conditions, and a good measure of anti-peasant sentiment, he claims, decided the Party on a "collision course" with the peasantry which was to escalate massively and unpredictably. See, Robert Conquest, *The Harvest of Sorrow: Soviet Collectivization and the Terror-Famine* (Oxford: Oxford University Press, 1986), 87-91. The findings of my own study are most compatible, however, with Lynne Viola's notion of state and peasantry as "two cultures at fatal variance with one another," and with her claim that "the state sought the eradication of peasant culture and autonomy, [and] the focused acculturation of the peasantry into the dominant culture" (a process she calls, "depeasantization." My findings also concur with Conquest's notion of the NEP era as one in which the confrontation between state and peasantry was a "stalemate." Lynne Viola, *Peasant Rebels under Stalin: Collectivization and the Culture of Peasant Resistance* (Oxford: Oxford University Press, 1996), 3.

national or class, gain. In the story of early Soviet forest conservation, in other words, we see a stark microcosm of—and gain new insight into—what Lynne Viola has called a "war of cultures," and Robert Conquest has dubbed a "stalemate," between state and peasantry. However construed, this tension was brutally resolved after 1928.[10] Early Soviet forest disputes, thus, probably *should* be set alongside other more well-known examples of peasant-state hostility, such as over grain requisitions during the Civil War, or regarding the persistence during the 1920s of peasant land repartition and the *mir*.

Second, an examination of forest conservation has at least some relevance to questions of the legitimacy and ultimate nature of the Bolshevik Revolution itself. I refer here to the old historiographical chestnut of whether October 1917 should be understood as a true social revolution, expressing the interests and will at least of a significant element or number of the Russian population, or as a simple seizure of power by self-interested parties (the Bolsheviks) for their own benefit.[11] Here, of course, forest history can provide only insights, not definitive answers—it is by its very nature far too narrow a field to do more. These caveats aside, the following can be argued. From the perspective of the peasantry, the Bolsheviks were an unwanted and even foreign intrusion into local life. Not only were rational, centrally-directed, and nationally-organized forestry totally rejected by the peasants, but the latter, far from uninterested in forest affairs themselves, possessed and articulated (via SR representatives, Land Committees, and otherwise) contrary—and even somewhat coherent—ideas of their own, involving local and unplanned use. While the Bolsheviks were forced to compromise with these somewhat, they never reconciled themselves to them. Viewed from this angle, then, October certainly seems more a seizure of power than a social revolution. Based on what has been argued throughout this study, it would be extremely difficult even to assert that the peasants, in spite of their objections, nonetheless reaped at least some tangible benefits from whatever

---

[10] Lynne Viola, *Peasant Rebels under* Stalin, 3 and passim. Robert Conquest, *The Harvest of* Sorrow, 87-91. Mark Von Hagen has also highlighted state-society tensions, in particular the visceral hatred of the Bolsheviks for peasant culture and life. See Mark Von Hagen, *Soldiers in the Proletarian Dictatorship: The Red Army and the Soviet Socialist State, 1917-1930* (Ithaca, NY: Cornell University Press, 1990).

[11] This particular historiographical terrain is clearly mapped out in Ronald Grigor Suny, "Revision and Retreat in the Historiography of 1917: Social History and Its Critics," *The Russian Review* 53 (1994): 165-82.

changes the Bolsheviks did manage to implement. It seems they did not.

On the other hand, it might be possible to see a more "representative revolution" by focusing not on the peasants but instead on the community of foresters, conservationists, and specialists. These—and the "consensus" they developed and promoted shortly prior to October, involving nationalization, "organic centralization," and scientific planning—did indeed find expression in Bolshevik forest legislation, as has been argued in some detail in chapters 4-6. The assumption of control over these changes by politicians, and the concomitant subordination of specialists to political authorities, however, muddies the waters, making simple extrapolations from foresters' plans to Bolshevik realities untenable.

Ultimately—in the forest domain—the first decade or so of the Bolshevik Revolution brought no obvious benefit to the peasantry, only slightly more to the majority of foresters, and more apparent harm than good to Russia's forests themselves.

Selected Bibliography

I. Archival Sources

Archive of Moscow State University, Moscow
Fond 200 (General history of conservation in Russia and the Soviet Union).

Bakhmetev Archives, Columbia University, New York. (Manuscript Collections of V. P. Semenov-tian-Shanskii and V. I. Vernadskii).

Gosudarstvennyi arkhiv rossiiskoi federatsii. (GARF, Moscow)
Fond A-358 (the hunting economy).

Fonds 2306, 2307 (*Narkompros* [Peoples' Commissariat of Education]).

Fond 5520 (*Soiuz lesovodov* [The Foresters' Union]).

II. Publised Document Collections

Iakovlev, N. I., compiler. *Sbornik uzakonenii o sberezhenii lesov chastnykh i obshchestvennykh v guberniiakh evropeiskoi rossii i tsarstva pol'skogo.* Saint Petersburg, 1902.

Organizatsiia nauki v pervye gody Sovetskoi Vlasti (1917-25). Sbornik dokumentov. Otv. red.: K.V. Ostrovitianov. Leningrad: "Nauka," Leningradskoe Otdelenie, 1968.

Rossiia. *Polnoe sobranie zakonov Rossiiskoi imperii.* (Selected years, 1770-1880).

_____. *Sobranie uzakonenii i rasporiazhenii rossiiskogo pravitel'stva* (Selected years, 1880-92).

RSFSR. *Dekrety Sovetskoi vlasti.* Moscow: Gosudarstvennoe izdatel'stvo politicheskoi literatury, 1957-1990.

_____. Narodnyi komissariat po prosveshcheniiu. *Dekrety, instruktsii i rasporiazheniia Narkomprosa po okhrane prirody.* Moscow and Leningrad: Glavnauka, 1929.

_____. *Sobranie uzakonenii i rasporiazhenii raboche-krestianskogo pravitel'stva.* Moscow: Narodnyi komissariat iustitsii, 1917-1925.

III. Conference Proceedings, Stenographic Records, Official Publications

Ditmar, B. P., and S.A. Severtsov, eds. Vserossiiskii s"ezd po okhrane prirody. Pervyi. *Trudy*. Moscow, 1930.

Imperatorskoe Russkoe geograficheskoe obshchestvo. Postoiannaia prirodookhranitel'naia komissiia. *Mirovaia okhrana prirody*. (Published as Vypusk 2 of the Commission's Occasional Papers). Ed. V. A. Dubianskii. Trans. from the French by E. Erimina. Petrograd, 1915.Rossiia. Gosudarstvennyi sovet. *Stenograficheskie otchety*. Saint Petersburg: Gosudarstvennaia tipografiia, 1906-1917.

RSFSR. Glavnyi Lesnoi Komitet. Upravlenie lesozagotovok. *God raboty po snabzheniiu zheleznykh dorog toplivom*. Gosudarstvennoe izdatel'stvo, 1920.

_____. Ministerstvo finansov, *Narodnoe khoziaistvo v 1914 g*. Petrograd, 1916.

_____. *Narodnyi komissariat zemledeliia, otchet, 1924-1925*. Moscow: Novaia derevnia, 1926.

_____. *Protokoly zasedanii vserossiiskogo tsentral'nogo ispolnitel'nogo komiteta* Stenograficheskii otchet. Moscow: Gosudarstvennoe izdatel'stvo, 1918-27. (Since numerous variant citations exist for this source—depending on the year, meeting, and so on—full citations are given for every usage within the relevant footnotes).

_____. *Vserossiiskii tsentral'nyi ispolnitel'nyi komitet*. Stenograficheskii otchet. Izdanie VtsIK, 1923-1929.

_____. *Vserossiiskii tsentral'nyi ispolnitel'nyi komitet sovetov rabochykh, krestianskykh, kazach'ikh i krasnoarmeiskykh* deputatov. Moscow: Izdatel'stvo VtsIK, 1919.

_____. *XVI S'ezd vsesoiuznoi kommunisticheskoi partii (b) 26 iiunia - 13 iiulia 1930,* Stenograficheskii otchet. Partizdat TsK VKP[b], 1935.

Vsekarel'skii s"ezd predstavitelei trudiashchikhsia karel, 1-3 iuliia 1920 [and] *Vsekarel'skii s"ezd sovetov rabochykh, krestianskykh i krasnoarmeiskykh deputatov, 11-18 fevralia 1921. Protokoly*. Petrozavodsk: "Karelia," 1990.

Vserossiiskaia sel'sko-khoziastvennaia i kustarno-promyshlennaia vystavka 1923 g. *Almanakh*. Moscow, 1923.

Vserossiiskii s"ezd okhotnikov v Moskve s 17-ogo do 25-ogo noiabria 1909 g. Vtoroi. *Trudy*. Moscow: 1910.

Zograf, N. Iu., ed. Vserossiiskii iubileinyi akklimatizatsionnyi s"ezd. *Trudy*. Vypusk 1. Obshchie sobraniia s"ezda. Moscow, 1909.

IV. Journals (1905-1931)

Biulleten' Narkomzema R.S.F.S.R..

Biulleten' tsentral'nogo upravleniia lesnoi promyshlennosti (glavleskoma) i tsentral'nogo komiteta profsouiza derevoobdelchikov

Estestvoznanie v shkole (Obshchestvo rasprostraneniia estestvenno-nauchnogo obrazovaniia).

Estestvoznanie v sovetskoi shkole (Organ uchebno-metodicheskoi sektsii Narkomprosa).

Les

Lesa respubliki

Lesnaia kooperatsiia

Lesnaia zhizn' i khoziaistvo

Lesnoe khoziaistvo

Lesnoe khoziaistvo i lesnaia promyshlennost'

Lesnoe khoziaistvo, lesopromyshlennost' i toplivo

Lesnoi dukh

Lesnoi spetsialist

Lesnoi zhurnal

Lesopromyshlennoe delo

Lesopromyshlennyi vestnik

Lesovod

Okhotnik: Illustrirovannyi zhurnal okhoty i rybolovstva

Okhrana prirody

Okhrana prirody i zapovednoe delo v SSSR

Priroda

V. General Reference Works

Biograficheskii slovar' deiatelei estestvoznaniia i teкhniki. Moscow, 1958-59.

Bol'shaia Entsiklopediia. Brokgaus edition. Moscow: Brokgaus, 1900.

Bol'shaia Sovetskaia Entsiklopediia. Moscow, 1920, 1950.

Konn, Tania, ed. *Soviet Studies Guide*. Glasgow, U.K. 1992.

Polnaia entsiklopediia russkogo sel'skogo khoziaistva. Saint Petersburg, 1903.

VI: Books and Articles

Abramovich. *O krestianskykh servitutakh*. 1895.

Al'fa. "Ne priznaiushchye kommunisma." *Lesa respubliki* 16 (1918): 863-66.

Algvere, Karl. V. *Forest Economy in the USSR*. Stockholm: Royal College of Forestry, 1966.

Anuchin, D. N. *Okhrana pamiatnikov prirody*. Saint Petersburg, 1914.

Astanin, L. P., and K. N. Blagosklonov. *Conservation of Nature*. Moscow: Progress Publishers, 1983.

Babeau, Hubert. *La Forête de Chizé*. Poitiers: Étude geographique. Université de Poitiers, 1973.

Bailes, Kendall E. *Environmental History: Critical Issues in Comparative Perspective*. Lanham, MD: University Press of America, 1985.

_____. *Science and Russian Culture in an Age of Revolutions: Vladimir Vernadsky and his Scientific School*. Bloomington: Indiana University Press, 1990.

Bannikov, A.G. *Po zapovednikam Sovetskogo Soiuza*. Moscow: Izdatel'stsvo Mysl', 1974.

Barr, B. and K. Braden. *The Disappearing Russian Forest: A Dilemma in Soviet Resource Management*. London: Hutchinson; and Totowa, NJ: Rowman and Littlefield, 1988.

Bergkholtsas, I. *Baltika: Okhrana prirodnoi sredy i* pravo. Riga: "Avots," 1986.

"Beseda o lese i ego znachenie v zhizni naroda." *Lesnaia zhizn' i khoziaistvo* 2 (1917): 5-11.

Blagosklonov, K. N., and A.A. Inozemstev, and V. N. Tikhomirov. *Okhrana Prirody.* Moscow: Vysshaia Shkola, 1967.

Boguslavskii, G. A. "Iz istorii okhrany prirody v pervye gody Sovetskoi vlasti." In *Stroitel'stvo Sovetskogo gosudarstva: Sbornik statei k 70-letiiu doktora istoricheskykh nauk, professora E. B. Genkinoi.* Izdatel'stvo "Nauka," 1972.

Bonhomme, Brian. "A Revolution in the Forests? Forest Conservation in Soviet Russia, 1917-1925." *Environmental History* 7:3 (July 2002): 411-34.

"Bor'ba protiv Osnovnogo Zakona." *Lesa respubliki* 8-9 (1918): 461-66.

Boreiko, V. E. *Belye piatna istorii prirodookhrany: Tom vtoroi, SSSR, Rossiia, Ukraina.* Seria: Istoriia okhrany prirody, vypusk 7. Kiev: Kievskii ekologo-kul'turnyi tsentr, Tsentr okhrany dikoi prirody SoES, 1996.

_____. "Plakala Sasha kak les vyrubili." In Boreiko, V. E. *Belye piatna istorii prirodookhrany: Tom vtoroi, SSSR, Rossiia, Ukraina.* Seria: Istoriia okhrany prirody, vypusk 7. Kiev: Kievskii ekologo-kul'turnyi tsentr, Tsentr okhrany dikoi prirody SoES, 1996, 5-64.

Bortkevich, V. M. "Moskovskii Oblastnoi S"ezd Soiuza Lesovodov." *Lesnoi zhurnal* 3-5 (1918): 113-19.

Bowler, Peter. *The Norton History of the Environmental Sciences.* New York: W. W. Norton, 1993.

Bramwell, Anna. *Ecology in the 20th Century: A History.* New Haven, CT: Yale University Press, 1989.

Brown, John Croumbie. *Forestry in the Mining Districts of the Ural Mountains in Eastern Russia.* Edinburgh: Oliver and Boyd, etc., 1884.

Bryzgalin, G. A., and Zakharov, B. A. *Chto takoe natisional'nye parki i dlia chego oni uchrezhdaiutsa?* Khar'kov, 1919.

Chekhov, Anton. *Uncle Vanya.* In *Anton Chekhov's Plays, A Norton Critical Edition.* Trans. and Ed. Eugene K. Bristow. New York: W. W. Norton, 1977.

Chertov, L. G. "Ekonomicheskaia otsenka lesov Evropeiskogo severa SSSR i prirodnykh uslovii ikh eksploatatsii." In *Geograficheskii sbornik XI: Ekonomicheskaia geografiia.* Moscow and Leningrad: Izdatel'stvo Akademii nauk SSSR, 1957, 21-49.

Clapp, B. W. *An Environmental History of Great Britain.* New York: Longman, 1994.

Courtois Stephane, et al. *The Black Book of Communism: Crimes, Terror, Repression.* Cambridge, MA: Harvard University Press, 1999.

Danishevskii, K. "Novyi kurs." *Biulleten' tsentral'nogo upravleniia lesnoi promyshlennosti (glavleskoma) i tsentral'nogo komtiteta profsoiuza derevoobdelchikov* 10-11

(1922), 1.

Davydova, M., and V. Koshevoi. *Nature Reserves in the USSR*. Moscow: Progress Publishers, 1989.

Debardeleben, Joan. *The Environment and Marxism-Leninism*. Boulder, CO: Westview Press, 1985.

_____, ed. *To Breath Free: Eastern Europe's Environmental Crisis*. Washington: The Woodrow Wilson Center Press, 1991.

Denis'ev, A. "O bor'be s samovol'nymi porubkami lesa." *Lesnoe khoziaistvo i lesnaia promyshlennost'* 1 (1928), 53-58.

Dergunov, Nikolai. *Okhrana i privlechenie ptits*. Moscow: Izdatel'stvo "Rabotnik prosveshchenia," 1927.

Dobrokhotov, Vladimir Ivanovich. *Astrakhanskii gosudarstvennyi zapovednik*. 2d. ed. Moscow, 1940.

Dokhman, G. I. *Istoriia geobotanika v Rossii*. Moscow: Nauka, 1973.

"Doklad Raionnomu S"ezdu Lesnichikh IV raiona tambovskoi gubernii." *Lesnaia zhizn' i khoziaistvo* 1 (1917): 9-16.

Elchibegoff, Ivan M. *Georgii Fedorovich Morozov: Father of the Russian Forestry Science*. Self-published, Washington, 1967.

Ermolov, A. S. *Sovremennaia pozharnaia epidemiia v Rossii*. 1910.

Faleev, N. I.. "Chto sdelano i chto predstoit' sdelat'." *Lesa respubliki* 8-9 (1918): 467-78.

_____. "Godovshchina revoliutsii." *Lesa respubliki* 16 (1918): 835-39.

_____. "Izmenenie polozheniia pravitel'stvennykh lesnykh spetsialistov." *Lesa respubliki* 3 (1918): 165-68.

_____. "Kapitalisticheskye osnovy lesnogo khoziaistva." In N. I. Faleev, ed., *Osnovnoi zakon o lesakh, izdanie neofitsial'noe*. Moscow: Izdanie Narkomzema, 1918, 1-20.

_____., ed. *Osnovnoi zakon o lesakh, izdanie neofitsial'noe*. Moscow: Izdanie Narkomzema, 1918.

_____. "Otchuzhdenie lesov." In N. I. Faleev, ed., *Osnovnoi zakon o lesakh, izdanie neofitsial'noe*. Moscow: Izdanie Narkomzema, 1918, 62-82.

_____. "Otmena lesnoi sobstvennosti." In N. I. Faleev, ed., *Osnovnoi zakon o lesakh, izdanie neofitsial'noe*. Moscow: Izdanie Narkomzema, 1918, 39-61.

_____. "Programma i sistema osnovnogo zakona o lesakh." In N. I. Faleev, ed., *Osnovnoi zakon o lesakh, izdanie neofitsial'noe*. Moscow: Izdanie Narkomzema, 1918, 30-38.

_____. "Starye lesnye osnovy," *Lesa respubliki* 5-7 (1918). 265-70.

Faleev, N. I., and T. Nekhoroshev, and D. Sazhin. "Lesnye organy," *Lesa respubliki* 5-7 (1918): 331-42.

Feoktistov, K. Kh. *Polozhenie ob okhrane lesov respubliki i instruktsii lesnoi strazhi po okhrane prirody*. Leningrad, 1924.

Fernow, Bernard. *A Brief History of Forestry in Europe, the United States and Other Countries*. Toronto: University of Toronto Press, 1911.

Feshbach, Murray. *Ecological Disaster: Cleaning up the Hidden Legacy of the Soviet Regime*. New York: Twentieth-Century Fund Press, 1995.

_____., ed. *Environmental and Health Atlas of Russia*. Moscow: PAIMS Publishing House, 1995. Distributed by the Center for Post-Soviet Studies, 2 Wisconsin Circle, Ste. 410, Chevy Chase, Maryland 20815, USA.

_____. and Alfred Friendly, Jr. *Ecocide in the USSR: Health and Nature Under Siege*. New York: Basic Books, 1992.

Fok, A. "Zashchitnye lesa." *Lesa respubliki* 5-7 (1918): 323.

Fol'mar, Georg. "Sovremmenoe sostoianie voprosa o lesookhranenii." *Lesnoi zhurnal* 7 (1907), 1098-1104.

Gare, Arran. "Soviet Environmentalism: The Path Not Taken." In *Capitalism, Nature, Socialism: A Journal of Socialist Ecology* 4 (1993): 69-88.

Genko, N. K. *K statistike lesov Evropeiskoi Rossii*. Saint Petersburg, 1888.

Geograficheskii sbornik XI: Ekonomicheskaia geografiia. Moscow and Leningrad: Izdatel'stvo Akademii nauk SSSR, 1957.

Gershenkron, Alexander. *Economic Backwardness in Historical Perspective: A Book of Essays*. Cambridge, MA: Belknap Press of Harvard University Press, 1962.

Gershman, I. "Ocherk istorii lesovladeniia, lesnoi sobstvennosti i lesnoi politiki v Rossii." *Lesnoi zhurnal* 3-4 (1911): 493-518.

Giriaev, D. and A. Savel'ev. "Kratkii ocherk istorii lesnogo khoziastva Rossii." *Lesnoe khoziastvo* 1 (1992): 1-10.

Gladkov, N. A. *Okhrany prirody v pervye gody Sovetskoi vlasti (po materialam leninskikh dekretov)*. Moscow: Izdatel'stvo moskovskogo universiteta, 1972.

Goldman, Marshall. *The Spoils of Progress: Environmental Pollution in the Soviet Union.* Cambridge, MA: MIT Press, 1972.

Graham, Loren. *Science and Philosophy in the Soviet Union.* New York: Knopf, 1972.

_____. *Science in Russia and the Soviet Union: A Short History.* New York: Cambridge University Press, 1993.

_____. *The Soviet Academy of Sciences and the Communist Party 1927*-32. New Jersey: Princeton University Press, 1967.

Gratsianov, V. I. *Kratkii ocherk istorii Imperatorskogo russkogo obshchestvo akklimatizatsii zhivotnikh i rastenii za piat'desiat let sushchestvovaniia, 1857*-1907. Moscow, 1907.

Grossgeim, A. A. *Rol' zapovednikov v khoziaistve strany.* Baku, 1929.

Grove, Richard H. *Green Imperialism: Colonial Expansion, Tropical Island Edens, and the Origins of Environmentalism 1600*-1860. New York: Cambridge University Press, 1995.

Gurov, P. Ia., ed. *Lesnoi kodeks RSFSR (v voprosakh i otvetakh).* Moscow, 1924.

Heiss, L. *Askania-Nova, Animal Paradise in Russia: Adventures of the Falz-Fein Family.* London: Bodley Head, 1970.

"Istoriia zakona." *Lesa respubliki* 5-7 (1918): 397-404.

"Iz materialov lesnogo oblastnogo soveshchaniia, sostoiavshigosia v Petrograde 16-20 aprelia 1918 goda." *Lesnoi zhurnal* 3-5 (1918): 120-23.

Jancar-Webster, Barbara W. *Environmental Management in the Soviet Union and Yugoslavia: Structure and Regulation in Federal Communist States.* Duke University Press, 1987.

Joravsky, David. *Soviet Marxism and Natural Science, 1917-32.* New York: Columbia University Press, 1970.

Kabanov, G. "Itogi S"ezda i zadachi Soiuza." *Lesopromyshlennoe delo* (no number, June-July, 1922), 1-2.

Karimov, Alexei. "State, Navy and the Origin of Petrine Forest Cadastre." Published on-line by the Institute for the History of Science and Technology, Russian Academy of Sciences, Moscow, Russia. http://www.ihst.ru/~krm

K. B. "K voprosu o edinom lesnom organe." *Lesopromyshlennoe delo* 7-8 (1922): 5.

Kern, E. E. "Brestskii dogovor i russkye lesa." *Lesnoi zhurnal* 6-8 (1918): 143-53.

_____. *Les i ego znachenie v prirode: Vvedenie k izucheniiu osnovnogo zakona a lesakh 27-ogo maia 1918 goda.* Petrograd, 1919.

Khutortsov, I. I. *V. I. Lenin ob okhrane prirody*. Maikop: Adygeiskoe otdelenie krasnodarskogo knizhnogo izdatel'stva, 1969.

Kliuchevsky, V. O. *A History of Russia*. Trans. C. G. Hogarth. Vol. I. New York: Russell and Russell, 1960.

Kliuchnikov, V. F. "Konkretnye zadachi lesookhraneniia." *Lesnoi zhurnal* 9-10 (1911): 1410-24.

Knipovich, B. N. *Ocherk Deiatel'nosti N[ar]K[om]Z[em]a za 3 goda (1917-1920)*.Gos-izdatel'stvo, 1920.

Knystautas, A. *The Natural History of the USSR*. London, UK: Century, 1987.

Komarov, Boris (pseudonym of Zev Wol'fson). *The Destruction of Nature in the Soviet Union*. New York: M. E. Sharpe, 1980.

"Kompententsiia lesnykh otdelov i lesnogo departamenta." *Lesa respubliki* 2 (1918): 119-26.

"Kontrol' federal'noi vlasti." *Lesa respubliki* 5-7 (1918): 343-46.

Kovalin, D. T., ed. *Lesokhoziaistvennaia i lesomeliorativnaia nauka v SSSR*. Moscow: "Lesnaia promyshlennost'." 1967.

Kozlov, P. K. *Askania-Nova*. Petrograd, 1915.

Krupskaia, N. K. *Iz vospominanii o V. I. Lenine. Sbornik Vospominanii o V. I. Lenine*. Vol. 1. Moscow, 1956.

Kurazhovskii, Iu. N. *Iz istorii organizatsii okhrany prirody v astrakhanskom krae: Materialy i deiatel'nosti V. I. Lenina i ego soratnikov*. Astrakhan, 1959.

_____. *Vladimir Il'ich Lenin i priroda*. Astrakhan: Nizhe-volzhskoe knizhnoe izdatel'stvo, 1969.

_____. *Zapovednoe delo v SSSR*. Rostov-on-Don, 1977.

_____., compiler. *Iz istorii okhrany prirody v Astrakhanskom krae. Materialy i deiatel'nosti V. I. Lenina i ego soratnikov v oblasti okhrany prirody*. Astrakhan, 1958.

Kuznetsov, N. A. "Propaganda osnovnogo zakona o lesakh." *Lesa respubliki* 16 (1918): 870-72.

_____. "Ves'ma vazhnyi pamiatnik prirody—'Les na reke Vorskle'." *Priroda* 5-6 (1926): 97-99.

Lavrov, I. "Mezhdu dvumia s"ezdami." *Biulleten' tsentral'nogo upravleniia lesnoi promyshlennosti (glavleskoma) i tsentral'nogo komiteta profsoiuza derevoobdelchikov* 10-11

(1922), 12.

Lemeshev, M. Ia., *Bureaucrats in Power: Ecological Collapse*. Trans. Sergei Chulaki Moscow: Progress Publishers, 1990.

Lenin, Vladimir Il'ich. *Collected Works* (English). Trans. Bernard Isaacs, 3rd printing. Moscow: Progress Publishers, 1977.

Leonov, Leonid. *The Russian Forest*. Trans. Bernard Isaacs. Moscow: Progress Publishers, 1966.

"Lesnaia belaia gvardiia." *Lesa respubliki* 14-15 (1918): 749-51.

"Lesnaia sektsiia: Itogi rabot lesnoi sektsii." *Lesa respubliki* 17-19 (1918-1919): 899-901.

"Lesnaia sektsiia na pervom Kievskom Sel'skokhoziaistvennom S"ezde i printsip obshchego lesookhraneniia." *Lesnoi zhurnal*, 1-2 (1910), 1365-78.

"Lesnaia sektsiia na pervom Kievskom Sel'skokhoziaistvennom S"ezde i printsip obshchego lesookhraneniia" (continuation). *Lesnoi zhurnal*, 9-10 (1913): 1414-37.

Lesnoe khoziaistvo SSSR za 50 let, 1917-1967. Moscow: "Lesnaia promyshlennost," 1967.

"Lesnoi kodeks." *Lesnaia kooperatsiia* 10-11 (1923): 7-9.

Lesokhoziaistvennaia i lesomeliorativnaia naiuka v SSSR (1917-1967 gg). Moscow: "Lesnaia promyshlennost'," 1967.

"Lesovody-kommunisty." *Lesa respubliki* 14-15 (1918): 751-54.

"Levye s.-r. i lesnoi zakon." *Lesa respubliki*, 8-9 (1918): 481-88.

Lewis, Martin W. *Green Delusions: An Environmentalist Critique of Radical Green Environmentalism*. Durham: Duke University, 1992.

Lydolph, P. E. *Climates of the Soviet Union* (World Survey of Climatology, vol. 7). Amsterdam, Netherlands: Elsevier.

Maier, V. "Lesnoi Kodeks." *Lesopromyshlennoe delo* 13-14 (1923): 1-2.

Makarov, V. N. *Okhrana prirody v SSSR*. Moscow: Gosudarstvennoe izdatel'stvo kulturno-prosvetitel'noi literatury, 1947, 1949.

_____. *Zapovedniki SSSR*. Moscow: Sel'khozgiz, 1940.

Mashkov, B., and G. Rusanov. *The Astrakhan Reservation*. Moscow: Planeta, 1972.

Massey-Stewart, John, *The Nature of Russia*. London: Boxtree, in association with Anglia Television,

1992.

_____. *The Soviet Environment*. Cambridge: Cambridge University Press, 1992.

_____., ed. *The Soviet Environment: Problems, Policies and Politics*. Cambridge: Cambridge University Press, 1991.

Maynard, Sir John. *The Russian Peasant and Other Studies*. New York: Collier Books, 1942.

Mellor, R. E. H. *Geography of the USSR*. London, UK: Macmillan, 1964.

Mnatsakanian, Ruben A. *Environmental Legacy of the Former Soviet Republics*. Edinburgh, UK., 1992.

Morozov, Georgii Fedorovich. *Uchenie o lese*. Saint Petersburg, 1912.

Morozov, P. "Chto delat'?" *Lesnoe khoziaistvo, lesopromyshlennost' i toplivo,* 9-10 (1925): 5-11.

Ne-Lesovod. "Lesnaia revoliutsiia." *Lesa respubliki* 16 (1918): 848-51.

_____. "Okhrana lesa." *Lesa respubliki* 5-7 (1918): 321-22.

_____. "S"ezd lesnykh komissarov." *Lesa respubliki* 3 (1918): 135-38.

Novikov. G. A. *Ocherki po istorii ekologii*. Moscow, 1970.

Novitskyi. *Servituty i obiazatel'nye razverstanye v iugo-zapadnom krae*, 1881.

"Novoe Lesnoe Obshchestvo." *Les* 5 (1917): 11-15.

"O deistvitel'nykh razmerakh russkogo dovoennogo lesnogo eksporta." *Lesnoe khoziaistvo i lesnaia promyshlennost'*, 8-9 (1928): 49-55.

Okhotnich'e khoziaistvo R.S.F.S.R. Pod obshchei redaktsiei V. V. Dezhina. Moscow: "Lesnaia promyshlennost'," 1978.

Okhotnich'e khoziaistvo SSSR. Pod redakstiei N. N. Grakova. Moscow: "Lesnaia promyshlennost'," 1973.

"Osnovnoi zakon o lesakh." *Lesopromyshlennoe delo* 3-4 (1922): 15-16.

Parsons, Howard L. *Marx and Engels on Ecology*. Westpoint CT: Greenwood Press, 1977.

Peterson, D. J. *Troubled Lands: The Legacy of Soviet Environmental Destruction*. Boulder, CO: Westview Press, 1993.

Petrov, V. V. *Ecological Legislation of Russia*. Moscow, 1995.

Piatakov, G. "Kontsessionaia politika v lesnom dele." *Lesopromyshlennoe delo* 5-6 (1923): 1-3.

_____. "Osnovnye zadachi lesnoi promyshlennosti." *Lesopromyshlennoe delo* 1-2 (1923), 1.

Pod"iapol'skii, N. N. "Vladimir Il'ich i okhrana prirody." *Okhrana prirody* 2 (1929): 35-38.

"Poriadok naznacheniia lesotekhnicheskogo personala." *Lesa respubliki* 2 (1918): 114.

"Postanovlenye, vynesennye vneocherednym Vserossiiskim S"ezdom lesovladel'tsev i lesokhoziaev dlia obsuzdeniia lesookhranitel'nogo zakona." Held in St. Petersburg 23-27 January, 1911. *Lesnoi zhurnal* 3-4 (1911), 377-89.

"Protokol' 9-go ocherednogo zasedaniia S. Peterburgskogo lesnogo obshchestva, 26 November 1905." *Lesnoi zhurnal* 2 (1906): 188-226.

Pryde, Philip R. *Conservation in the Soviet Union*. Cambridge: Cambridge University Press, 1972.

_____. *Environmental Management in the Soviet Union*. Cambridge: Cambridge University Press, 1991.

_____. *Environmental Resources and Constraints in the Former Soviet Republics*. Boulder, CO and Oxford, UK: Westview Press, 1995.

Purcell, Carrol W., ed. *From Conservation to Ecology: The Development of Environmental Concern*. New York: Thomas Y. Crowell, 1973.

"Raboty Tsentra." *Lesa respubliki* 8-9 (1918): 477-82.

Radkey, Oliver H. *The Election to the Russian Constituent Assembly of* 1917. Cambridge, MA: Harvard University Press, 1950.

_____. *The Unknown Civil War in Soviet Russia: A Study of the Green Movement in the Tambov Region, 1920*-1921. Stanford, CA: Hoover Institution Press, 1976.

Radzinsky, Edvard. *Stalin: The First In-Depth Biography Based on Explosive New Documents from Russia's Secret Archives*. Trans. H. T. Willets. New York, etc.: Doubleday, 1996.

"Rezoliutsii s"ezda." *Lesa respubliki* 17-19 (1918-1919), 907-10.

"Rezoliutsii XI vserossiiskogo s"ezda lesovladel'stev i lesokhoziaev v g. Tule." *Lesnoi zhurnal* 1-2 (1910): 4-10.

Riasanovsky, Nicholas V. *A History of Russia*, 4th edition. New York and Oxford: Oxford

University Press, 1984.

Rostovstev, V. "Lesokul'turnye raboty v Kalenkovicheskoi dache, Kalenkovicheskogo lesnichestva, Minskoi guburnii." *Lesnoi zhurnal* 8 (1913): 1206.

Sazhin, D. K. "27-fevralia. 27-ianvaria. 27-maia." *Lesa Respubliki* 16 (1918): 839-46.

Schapiro, Leonard B. *The Communist Party of the Soviet Union,* 2nd ed. New York: Random House, 1970.

_____. *The Origin of the Communist Autocracy; Political Opposition in the Soviet State: First Phase, 1917-1922.* Cambridge, MA: Harvard Univ. Press, 1965.

Semenov-Tian-Shanskii, A. P. "Svobodnaia priroda, kak velikii zhivoi muzei, trebuet neotlozhnykh mer ograzhdeniia." *Priroda* 4-6 (1919): 199-216

Serebrennikov, P. "O novom lesnom kodekse." *Lesopromyshlennoe delo* 17-18 (1923), 9-10.

_____. "Zadachi gosudarstvennogo lesoustroistva." *Lesnoi zhurnal* 1-2 (1918): 8-13.

Service, Robert. *Lenin A Political Life*, 2nd edition. Bloomington, IN: Indiana University Press, 1995.

"S"ezd lesnykh komissarov." Stenogram, *Lesa respubliki* 8-9 (1918): 525-38.

Shaposhnikov, Lev Konstantinovich. *Okhrana prirody v SSSR.* Moscow, 1961.

Shtil'mark, F. R. "The Evolution of Concepts about the Preservation of Nature in Soviet Literature." Trans. Roberta Reeder. In *Journal of the History of Biology* 25:3 (1992): 429-47.

Singleton, F., ed. *Environmental Problems in the Soviet Union and Eastern Europe* Boulder, CO: Lynne Rienner, 1987.

Solov'ev, V. "Krest'ianstvo i lesnye rabotniki." *Lesovod* 4-5 (1924): 9-11.

_____. "Les i krest'ianstvo." *Lesovod* 2-3 (1924), 13-17.

"Soiuz lesovodov i lesnykh tekhnikov." *Les* 5 (1917): 24-25.

Stalin, Joseph Vissarionovich. *Voprosy Leninisma.* 11th edition. Gosudarstvennoe izdatel'stvo politicheskoi literatury, 1953.

Struchkov, Anton Iu. "Nature Protection as Moral Duty: The Ethical Trend in the Russian Conservation Movement." In *Journal of the History of Biology* 25:3 (1992): 413-28.

Sukachev. "Krymskii Gosudarstvennyi Lesnoi Zapovednik." *Lesovod* 2-3 (1924), 27-29.

Suny, Ronald Grigor. "Revision and Retreat in the Historiography of 1917: Social History and Its

Critics." *The Russian Review* 53 (1994): 165-82.

Teplyakov, V. K., et al., *A History of Russian Forestry and its Leaders*. Pullman, WA: Department of Natural Resource Sciences and Cooperative Extension, Washington State University, 1998.

Tertius. "Revoliutsiia i mesta." *Lesa respubliki* 16 (1918): 853-57.

Timonov, V. E. "Okhrana prirody pri inzhenernykh rabotakh." *Priroda* 1-2 (1922): 71-86.

"Tsentral'noe Upravlenie Lesov respubliki." *Lesa respubliki* 5-7 (1918): 341-44.

Tucker, Robert. *Stalin as Revolutionary*. New York: Norton, 1973.

Tumarkin,Nina. *Lenin Lives! The Lenin Cult in Soviet Russia*. Cambridge, MA: Harvard University Press, 1983.

Turnbull, Mildred M. *Soviet Environmental Policies and Practices: The Most Critical Investment*. Aldershot, Hants. England, 1990.

"Uchet i mobilizatsiia lesnykh spetsialistov." *Lesa respubliki* 21-22 (1919): 1109-11.
Varenne de Fenille, P. C. M. *Mémoires sur l'administration forestière*. A Bourg, De l'impr. de C. C. G. Philipon, 1792.

Vasil'ev, P. V. *V. I. Lenin o lesakh i lesnom khoziaistve*. Izdatel'stvo "Lesnaia promyshlennost'," 1967. In D. T. Kovalin, ed., *Lesokhoziaistvennaia i lesomeliorativnaia nauka v SSSR*. Moscow: "Lesnaia promyshlennost'," 1967.

"V. I. Lenin kak okhotnik." *Okhotnik: Illustrirovannyi zhurnal okhoty i rybolovstva* 10 (1927): 31.

"V. I. Lenin na okhote." *Okhotnik: Illustrirovannyi zhurnal okhoty i rybolovstva* 5 (1928): 33.

Vinogradov-Nikitin, P. "Okhrana pamiatnikov prirody." *Lesnoi zhurnal*, 1-2 (1910): 146-54.

Vladimir Il'ich Lenin i priroda: Beseda dlia shkol'nikov. Irkutsk: Irkustkaia oblastnaia stantsiia iunykh naturalistov, 1980.

Volkogonov, Dmitri. *Lenin: A New Biography*. The Free Press, 1994.

Von Hagen, Mark. *Soldiers in the Proletarian Dictatorship: The Red Army and the Soviet Socialist State, 1917-1930*. Ithaca, NY: Cornell University Press, 1990.

"Vospominaniia sputnikov po okhote V. I. Lenina-Ul'ianova." *Okhotnik: Illustrirovannyi zhurnal okhoty i rybolovstva* 10 (1927): 32.

V pomoshch' zhigariu: Pechnoe uglezhzhenie. Prilozhenie k zhurnaly. *Les* 2 (Feb. 1917), 3-47.

Vsekarel'skii s"ezd predstavitelei trudiashchikhsia karel, 1-3 iuliia 1920 [and] *Vsekarel'skii s"ezd sovetov rabochykh, krestianskykh i krasnoarmeiskykh deputatov, 11-18 fevralia 1921. Protokoly*. (Published as a combined edition). Petrozavodsk: "Karelia," 1990.

"Vse lesnoe khoziaistvo—v VSNKh." *Lesnoi spetsialist* 5 (1930), 4.

"Vserossiiskii s"ezd zemel'nykh otdelov, komitetov bednoty i kommun. Protokoly." *Lesa respubliki* 17-19 (1918-1919), 941-56.

"Vserossiiskii s"ezd zemel'nykh otdelov, komitetov bednoty i kommun. Rezoliutsii." *Lesa respubliki* 17-19 (1919): 957-1005.

"Vserossiiskii Soiuz Lesovodov i Lesnykh Tekhnikov (Konduktorov)." *Lesnaia zhizn i khoziaistvo* 2 (1917): 45.

"Vtoroi delegatskii s"ezd Soiuza lesovodov." *Lesnoi zhurnal* 3-5 (1918): 336-45.

Vysotskii, Georgii. "Neskol'ko slov o lesopole: 1. Udobstva natsional'izatsii." *Lesnoi zhurnal* 1-2 (1918): 37-44.

Weiner, Douglas R. *A Little Corner of Freedom: Russian Nature Protection from Stalin to Gorbachev*. Berkeley and London: University of California Press, 1999.

_____. *Models of Nature: Ecology, Conservation and Cultural Revolution in Soviet Russia*. Bloomington: Indiana University Press, 1988. Reprinted 2000 by University of Pittsburgh Press.

Wolfson, Z. (A.k.a. Boris Kamarov), *The Geography of Survival: Ecology in the Past Soviet Era*. Armonk, NY, and London: M. E. Sharpe, 1994.

Yanitsky, O. *Russian Environmentalism: Leading Figures, Facts, Opinions*. Moscow: Mezhdunarodnye otnoshenye Publishing House, 1993.

Zaitsev, D. M. *Lesnoe Zakonodatel'stvo (Kommentaryi k proektu novogo Ustava Lesnogo). Doklad 1-omu Vserossiiskomu S"ezdu Predstavitelei Lesnogo Promyshlennosti i Torgovli*. St. Petersburg: Biuro po uchrezhdeniiu S"ezdov Predstavitelei Lesnogo Promyshlennosti i Torgovli,1914.

Zapovedniki SSSR. Sbornik statei. Ed., A. A. Solov'ev. Intro. V. N. Makarov. 2 vols Moscow: Gosizdat geograficheskoi iteratury, 1951.

Zhukov, P. I. *Leninskye printsipy okhrany prirody*. Minsk: Gosudarstvennyi komitet soveta ministrov Belorusskoi SSR po okhrane prirody, 1969.

"Zhurnal S"ezda Lesnichikh III-ego raiona Tambovskoi gubernii v 1916 godu." *Lesnaia zhizn' i khoziaistvo* 2 (1917): 19.

Ziegler, Charles E. *Environmental Policy in the USSR.* Amherst: University of Massachusetts Press, 1987.

Zile, Zigurds L. "Lenin's Contribution to Law: The Case of Protection and Preservation of the Natural Environment." In Bernard W. Eissenstat, ed., *Lenin and Leninism: State, Law and Society.* Lexington, MA.: D.C. Heath, 1971, 83-100.

Zirkle, Conway. *Evolution, Marxian Biology and the Social Scene*. Philadelphia, 1959.

Zubovskii, N. "Tsentrovye s. r. i lesa." *Lesa respubliki* 5-7 (1918): 291-94.

"Zverskoi ubiistvo 19 chelovek." *Lesovod* 1 (1924), 68-70.